Studies of Brain Function, Vol. 4

Coordinating Editor
V. Braitenberg, Tübingen

Editors
H. B. Barlow, Cambridge
E. Bizzi, Cambridge, USA
E. Florey, Konstanz
O.-J. Grüsser, Berlin-West
H. van der Loos, Lausanne

Heiko Braak

Architectonics of the Human Telencephalic Cortex

With 43 Figures

Springer-Verlag
Berlin Heidelberg New York 1980

Prof. Dr. HEIKO BRAAK
Dr. Senckenbergisches Anatomisches Institut
Theodor-Stern-Kai 7
6000 Frankfurt 70

ISBN 3-540-10312-0 Springer-Verlag Berlin Heidelberg New York
ISBN 0-387-10312-0 Springer-Verlag New York Heidelberg Berlin

Library of Congress Cataloging in Publication Data
Braak, Heiko, 1937–. Architectonics of the human telencephalic cortex. (Studies of brain function ; v. 4) Bibliography: p. Includes index. 1. Cerebral cortex. 2. Cytoarchitectonics. 3. Telencephalon. I. Title. II. Series. QM575.B72 611'.81 80-21134

This work is subject to copyright. All rights are reserved, whether the whole or part of the material is concerned, specifically those of translation, reprinting, re-use of illustrations, broadcasting, reproduction by photocopying machine or similar means, and storage in data banks. Under § 54 of the German Copyright law, where copies are made for other than private use, a fee is payable to the publisher, the amount of the fee to be determined by agreement with the publisher.

© by Springer-Verlag Berlin Heidelberg 1980.
Printed in Germany.

The use of registered names, trademarks, etc. in this publications does not imply, even in the absence of a specific statement, that such names are exempt from the relevant protective laws and regulations and therefore free for general use.

Offsetprinting and binding: Konrad Triltsch, Graphischer Betrieb, Würzburg
2131/3130-543210

Preface

This is a timely opus. Most of us now are too young to remember the unpleasant ring of a polemic between those who produced "hair-splitting" parcellations of the cortex (to paraphrase one of O. Vogt's favourite expressions) and those who saw the cortex as a homogeneous matrix sustaining the reverberations of EEG waves (to paraphrase Bailey and von Bonin). One camp accused the other of producing bogus preparations with a paint brush, and the other way around the accusation was that of poor eye-sight. Artefacts of various sorts were invoked to explain the opponent's error, ranging from perceptual effects (Mach bands crispening the areal borders) to poor fixation supposedly due to perfusion too soon (!) after death. I have heard most of this directly from the protagonists' mouths. The polemic was not resolved but it has mellowed with age and ultimately faded out. I was relieved to see that Professor Braak elegantly avoids discussion of an extrememist tenet, that of "hair-sharp" areal boundaries, which makes little sense in developmental biology and is irrelevant to neurophysiology. It was actually detrimental to cortical neuroanatomy, since its negation led to the idea that structurally distinct areas are not at all existent. Yet, nobody would deny the reality of five fingers on one hand even if the detailed assignment of every epidermal cell to one finger or another is obviously impossible.

Plenty of recent electrophysiological findings might have led one to postulate regional variations of the cortical wiring in patches of a few centimeters each, had they not been already described in papers on cortical "architectonics". With the electrophysiological analysis approaching the level of actual neuronal computation, we might be close to an explanation of areal variations of structure in terms of the different tasks performed.

Summer 1980
V. Braitenberg

Acknowledgements and Dedication

This study was partly worked out during a visit at Harvard University (Boston) in summer 1978. The author is greatly indebted to Professors N. Geschwind and T. Sabin for supplying the laboratory facilities, for their invaluable advice and constructive criticism. He would also like to express his appreciation to Professors I.T. Diamond (Durham), A. Galaburda (Boston), T. Kemper (Boston), S.L. Palay (Boston), D.N. Pandya (Bedford), A. Peters (Boston), C. West (Bedford), P. Yakovlev (Washington) for valuable discussion and invigorating comments. The studies were essentially brought forward by examining parts of the Yakovlev collection of serial sections through human brains (Kretschmann et al., 1979). The author is especially grateful for the privilege of having used this collection.

Completion of this book would have been impossible without the generous help and advice of many German colleagues and friends. The author is particularly indebted to Professors K. Fleischhauer (Bonn), R. Hassler (Frankfurt), H. Leonhardt (Kiel), H. Stephan (Frankfurt), I.R. Wolff (Göttingen), and K. Zilles (Kiel) for fruitful discussions. The author wishes to express his gratitude to the editors, in particular to Professor V. Braitenberg (Tübingen), for constant encouragement and critical reading of the manuscript. He is also indebted to his wife, Dr. Eva Braak, for her enduring patience, for valuable comments, and kind participation in proof reading.

The author wishes to thank Mr. R. Clemens for his help with the drawings and the Akademische Verlagsgesellschaft Geest and Porting (Leipzig) for kindly permitting use of illustrations which originally appeared in the *Zeitschrift für mikroskopisch-anatomische Forschung*. The skilful technical assistance of Mrs. S. Piontek (Kiel) is greatly acknowledged, as well as the further technical help of Mr. A. Zalkalns (Boston) and Mrs. R. Schneider (Frankfurt). The author is furterhmore indebted to Mr. G. Grow (Kiel) for improving the style and readability of the text, to Mrs. G. Schulze (Frankfurt) for her devoted secretarial services. He wishes to thank the publisher, Dr. K.F. Springer (Heidelberg) for accepting the manuscript and quickly guiding it through the various stages or production. Means for the research were kindly supplied by the Deutsche Forschungsgemeinschaft.

This work is dedicated to Professor W. Kirsche (Berlin) to whom neuroanatomy owes so many insights and fruitful concepts.

Summer 1980 H. Braak

Contents

1	Introduction	1
2	Types of Nerve Cells Forming the Telencephalic Cortex	3
2.1	Pyramidal Cells	3
2.2	Modified Pyramidal Cells	6
2.3	Non-Pyramidal Cells	10
3	The Three Standard Techniques Used in Architectonics	12
3.1	Preparations Stained for Nerve Cells (Nissl Preparations) as a Basis of Cytoarchitectonics	12
3.2	Preparations Stained for Myelin Sheaths as a Basis of Myeloarchitectonics	15
3.3	Preparations Stained for Lipofuscin Granules as a Basis of Pigmentoarchitectonics	18
4	The Main Subdivisons of the Telencephalic Cortex	24
5	The Allocortex	26
5.1	The Hippocampal Formation	26
5.1.1	The Fascia Dentata	26
5.1.2	The Cornu Ammonis	28
5.1.3	The Subiculum	30
5.2	The Presubiculum	34
5.2.1	The Proper Presubicular Subregion	36
5.2.2	The Parasubicular Subregion	36
5.2.3	The Transsubicular Subregion	37
5.3	The Entorhinal Region	37
5.3.1	The Proper Entorhinal Subregion	39
5.3.2	The Transentorhinal Subregion	42
6	The Proisocortex	49
6.1	The Retrosplenial Region	49
6.2	The Anterogenual Region	55

7	**The Mature Isocortex**	63
7.1	The Occipital Lobe	64
7.1.1	The Striate Area	64
7.1.2	The Parastriate Area	70
7.1.3	The Peristriate Region	73
7.2	The Temporal Lobe	74
7.2.1	The Temporal Granulous Core	76
7.2.2	The Temporal Progranulous Field	78
7.2.3	The Temporal Paragranulous Belt	78
7.2.4	The Temporal Magnopyramidal Region	80
7.2.5	The Temporal Uniteniate Region	84
7.3	The Parietal Lobe	85
7.3.1	The Parietal Granulous Core	86
7.3.2	The Parietal Paragranulous Belt	88
7.3.3	The Parietal Magnopyramidal Region	90
7.4	The Frontal Lobe	91
7.4.1	The Frontal Ganglionic Core	93
7.4.2	The Frontal Paraganglionic Belt	97
7.4.3	The Frontal Magnopyramidal Regions	98
7.4.3.1	The Inferofrontal Magnopyramidal Region	98
7.4.3.2	The Superofrontal Magnopyramidal Region	99
8	**Brain Maps**	104
9	**Notes on Techniques**	121
References		124
Subject Index		145

1 Introduction

As early as 1782, Gennari pointed to the fact that the telencephalic cortex is not a uniformly built structure. Since then, numerous additional findings have been published which allowed one to distinguish an increasing number of cortical areas. Analysis of the "architectonics" of the brain (as defined by Vogt and Vogt, 1937) means the study of all the subtle to clearly marked local variations caused by differences in the arrangement, the packing density, the size and the shape of the various components which contribute to the formation of the nerve tissue (Hassler, 1962).

The human cortex in particular permits distinction of a great number of structurally different fields. It is beyond the scope of the present text to give a full description of all the various areas which have been delineated already. The text limits itself to the description of only three standard techniques providing the bases for *cyto-, myelo-,* and *pigmentoarchitectonics,* to the illustration of only the most important characteristics revealed by the different preparations, and to the cortical mapping which has been achieved with the aid of these methods.

There are other methods that could be useful to the distinction of cortical areas. As is clear from Pfeifer (1940) the local variations of cortical blood vessels give valuable results (*angioarchitectonics*). But unfortunately, for technical reasons, the angioarchitectural analysis of the human brain cannot be considered a routine method (Duvernoy, 1979). Also varying amounts of enzymes in different locations is rarely used for parcellation of the human telencephalic cortex (*chemoarchitectonics,* Friede, 1966).

The study of *myelogenesis,* i.e., recognition of the sequence in which myelinated fibres appear underneath the different parts of the cortex, permits a basic subdivision (Flechsig, 1920, 1927). Analysis of *dendrogenesis,* i.e., recognition of the sequence in which the dendritic arbour of cortical pyramidal cells is established, requires particular experience of the observer and is furthermore bound to the more or less capricious silver impregnation techniques (de Crinis, 1934). Hence, the systematic part of the present text is deliberately based on the more easily accessible Nissl, myelin, and pigment preparations. The major emphasis is put on the description of cortical regions, the functional significance of which is relatively well known, such as the primary sensory and motor territories and their immediately adjoining areas. The relative neglect of "association" areas, which are particularly large in the human brain, seems reasonable since their further parcellation is difficult and requires particular training and experience. In fact, there is much controversy as to the subdivision of "association" territories and it is even unclear whether the concept of an "association" cortex can be upheld at all. Nevertheless, the "magnopyramidal areas", which for the most part belong to the "association" cortex, are treated in some detail since they can clearly be delineated with the aid of pigment preparations. These conspicuous fields become more and more the centre of interest of investigators concerned with the anatomical substrate of the "higher" functions of the human brain.

2 Types of Nerve Cells Forming the Telencephalic Cortex

On the basis of the Golgi method which stains individual neurons with their complete arborization most of the nerve cells which form the telencephalic cortex can be classified with either the pyramidal cells or the stellate cells (Fig. 1).

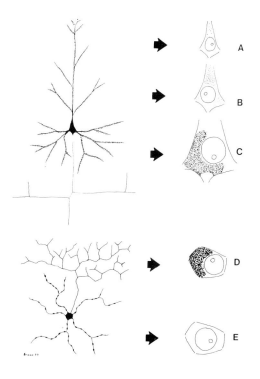

Fig. 1. Diagram showing the two main types of cortical nerve cells and their general pattern of pigmentation. The pyramids normally contain a loose distribution of finely grained pigment which can only weakly be stained by aldehydefuchsin (*A, B*). Only some varieties of large pyramids within the lower reaches of the third and the fifth cortical layer show a dense agglomeration of pigment (*C*) (pigment-laden IIIc- and Vb-pyramids, see Chap. 3.3). The stellate cells, in contrast, are either filled with coarse and intensely stained lipofuscin granules (*D*) or are devoid of pigment (*E*)

2.1 Pyramidal Cells

The pyramidal cells or pyramids derive their name from the appearence of their perikarya which in turn results from the polarized arrangement of the cellular processes (Campbell, 1905; Ramón y Cajal, 1909; von Economo and Koskinas, 1925; Jones and Powell, 1970a; Peters et al., 1970; Kirsche et al., 1973). In general, an apical dendrite is distinguishable from a variable number of less conspicuous processes arising from the base of the cell body.

The apical dendrite is more or less accurately oriented perpendicular to the cortical surface. It gives off a variable number of side branches and bursts into an umbel of terminal twigs within the molecular layer. The basal dendrites are generated from inferior tips of the pyramidal soma as usually slender processes which bifurcate repeatedly as they extend distally (Smit et al., 1972; Smit and Uylings, 1975; Uylings and Smit, 1975).

With the same general shape a scant number of pyramidal cells can be encountered in an abnormal orientation with their bases directed towards the cortical surface or obliquely disposed oriented at various angles. In general, the axon arises as in other pyramidal cells from the base of the cell body whence it initially runs in a direction opposite to that of the apical dendrite, but soon bends to pursue a straight course downwards (van der Loos, 1965; Globus and Scheibel, 1967; Meller et al., 1969; Kirsche et al., 1973) (Fig. 2 F).

Both the apical and the basal dendrites bear small spines which project from all sides (Koelliker, 1899; Ramón y Cajal, 1909; Ramón-Moliner, 1961; Jones and Powell, 1969b; Marın-Padilla et al., 1969; Kunz et al., 1972, 1974; Schierhorn et al., 1972a,b, 1973; Wenzel et al., 1973; Doedens et al., 1974, 1975; Schierhorn, 1978a,b,c). Most of these delicate appendages show a thread-like stem tilted at various angles and terminating in a rounded knob (Jacobson, 1967). Sessile spines by contrast are directly disposed upon the dendritic surface. Terminal ramifications of axons from various sources form synaptic contacts with the spines (Gray, 1959; Jones and Powell, 1969b; Peters and Kaiserman-Abramof, 1969; Le Vay, 1973). Spine density on the shafts of the apical dendrites is high on the large third-layer and fifth-layer pyramids and substantially less on the smaller pyramids of the other cortical laminae (Williams et al., 1979). The only regions almost barren of thorns are the soma and the proximal dendritic segments (Ramón y Cajal, 1891, 1909; Koelliker, 1896; Kirsche et al., 1973; Chan Palay et al., 1974). The spine-free zone of the dendritic shaft develops late in phylo- and ontogenesis. It is possibly the result of stretching of the cells after establishment of the axo-spinous synaptic contacts (Schierhorn et al., 1973; Doedens et al., 1974, 1975; Görne and Pfister, 1976).

The axon usually emerges from the base of the cell body by way of a broad cone-shaped axon hillock. Occasionally it may egress from a thick basal dendrite (Colonnier, 1967; Palay et al., 1968; Peters et al., 1968; Jones and Powell, 1969a; Peters and Kaiserman-Abramof, 1970; Kirsche et al., 1973). Radially oriented, it runs in a straight course towards the white matter and on its way gives off numerous collateral branches. These may run horizontally or ascend towards the surface. Collaterals take a straight course similar to the axon itself. They may run a fairly long distance before breaking up into terminal ramifications. Beyond the initial segment (Palay et al., 1968; Peters et al., 1968; E. Braak, 1980), the axon and at least some of its collateral branches are myelinated.

Onto- and phylogenesis of pyramidal cells show several similarities. After emergence of the axon the apical dendrite is the next process to be developed. Its terminal twigs in the molecular layer appear early. A varying number of basal dendrites follow. Side branches of the apical shaft appear last (Ramón y Cajal, 1906; Marin-Padilla, 1970a,b, 1971, 1972). Yet the skirt of basal dendrites and even more so the side branches given off from the stem of the apical dendrite become considerably more complex with phylogenetic advance. In the human telencephalic cortex a great number of apical dendrites do not extend up to the molecular layer and lack a terminal bouquet (H. Braak, 1976b; E. Braak, 1978b).

The spines develop late in ontogeny (Ramón y Cajal, 1909; Purpura, 1975). They project at first from the apical shaft and terminal tuft and thereafter from the basal processes and the side branches of the apical dendrite (Ramón y Cajal, 1909; Eayrs and Goodhead, 1959; Noback and Purpura, 1961; Poliakov, 1961, 1964/65, 1966; Schadé and van Groeningen, 1961; Purpura et al., 1964; Schadé et al., 1964a,b; Fox et al., 1966; Meller et al., 1968b, 1969; Morest, 1969; Molliver and van der Loos, 1969/70; Marin-Padilla, 1970a, 1971, 1972; Kirsche et al., 1973; Kirsche, 1974; Frotscher, 1975; H. Braak, 1976b; Görne and Pfister, 1976; Minkwitz, 1976c).

In routine preparations stained with a basic dye (Nissl preparations) only the roots of the dendrites are recognizable. In general, the stout and radially oriented apical dendrite can nevertheless be distinguished from the thinner basal ones. The apical cytoplasm tapers gradually and merges into the shaft. The cone-shaped dendritic roots can serve as a criterion for distinguishing pyramidal cells from stellate cells, the dendrites of which generally egress more abruptly from the cell body.

The soma contains Nissl granules of variable appearance; coarse and floccular in some types of pyramids or dispersed as fine and dust-like particles in others. Nissl granules can also be encountered in the proximal parts of the major dendrites.

Lipofuscin granules can be encountered in most of the nerve cell types of the adult human brain. The lipofuscin in nerve cells is different from that stored in glial cells or endothelial cells. There is also evidence for remarkable variations in the characteristics of the pigment from one type of nerve cell to another (H. Braak, 1978a). In general the amount of lipofuscin pigment stored in nerve cells increases slowly with advancing age (Mann and Yates, 1974; West, 1979). In pyramidal cells the pigment granules are generally randomly dispersed throughout the whole cell body and cannot be traced into the dendrites (Fig. 1 A, B). The axon is devoid of both the lipofuscin granules and the Nissl substance. Light-microscopically, also the axon hillock is sharply set off from the soma and appears free of pigment and Nissl substance (H. and E. Braak, 1976). Electron-microscopically, the axon hillock reveals a marked numerical reduction of pigment and basophilic material (E. Braak, 1980). The pigment within pyramidal cells is finely grained and can only faintly be tinged by aldehydefuchsin. The packing density of the lipofuscin granules depends on the type of the pyramidal cell and its location within the cortical layers (H. Braak, 1978a).

The rounded nucleus is usually in a slightly excentric position. Chromatin is present in only small amounts, often being confined to irregularly shaped patches under a distinct nuclear membrane. As a rule, in the adult brain the nucleus of pyramidal cells contains only one conspicuous nucleolus which often shows vacuoles (Schleicher et al., 1975; Zilles et al., 1976).

2.2 Modified Pyramidal Cells

Modified pyramidal cells show more or less thoroughgoing variations of their cellular processes which lead to considerable deviations from the standard appearance as described above.

The basal dendrites are not always of the same diametre and length. In places, one of these dendrites is of particular stoutness. The Meynert pyramids in layer Vb of the striate area often display such a main basal dendrite and only a tenuous and short apical dendrite (von Economo and Koskinas, 1925; Clark LeGros, 1942; Chan Palay et al., 1974; H. Braak, 1976b; E. Braak, 1978b; Palay, 1978) (Fig. 2 L).

The multiform layer of the isocortex contains a great number of nerve cells which emit only two stout dendrites each of approximately the same calibre and length. One of them is oriented perpendicular to the cortical surface, the other runs in various directions. These cells are therefore referred to as "a pair of compass cells" (de Crinis, 1933). The formation of only two main dendrites gives the cell body a triangular or rhombic contour. The triangular cells of Ramón y Cajal (1909) belong to this type of pyramidal cell. The axon arises frequently from the second main dendrite and descends straight into the white matter giving off collaterals on its way (Fig. 2 H, J, K).

Some types of pyramidal cells, for instance the tiny pyramids of the granular layer (IV), have apical dendrites which appear reduced to only a thread-like process. The axon heads towards the white matter but soon tapers off and in the author's experience cannot be followed further downwards. At a short distance from the cell body the axon gives off a relatively thick collateral which curves sharply upwards. Such cells can therefore be included among the cells with a local axon (Fig. 2 G) (Ramón y Cajal, 1900a, 1909; Lund, 1973; H. Braak, 1976b). It should be remembered that pyramidal cells in chronically isolated cortices show a similar pattern of their axons (Purpura and Housepian, 1961; Rutledge, 1978; Williams et al., 1979). Pyramids with only a short axon not extending into the white matter may also occur in the outer parts of the pyramidal layer (IIIab). Ablation and degeneration experiments buttress the assumption that the slender IIIab-pyramids establish intrinsic connections with axons terminating preferentially in the ganglionic layer (Jacobson, 1965; Spatz et al., 1970; Nauta et al., 1973; Levey and Jane, 1975; Lund and Boothe, 1975; Butler and Jane, 1977).

In other types of modified pyramidal cells the horizontal dendrites become the predominating cellular processes exceeding by far the delicate apical dendrite in length and diametre. The solitary cells of Cajal in layer IVb of the striate area for example show a cell body which appears as a flattened disk aligned parallel to the cortical surface. The solitary cells emit a straight descending axon which gives off a fair number of collaterals (Le Vay, 1973; Sanides and Sanides, 1974; H. Braak, 1976b; Tigges et al., 1977; E. Braak, 1978b) (Fig. 2 D).

The large to giant pyramids of layer Vb in motor cortices – the well-known Betz cells – often show a rather bizarre outline due to an enhanced number of primary dendrites. These issue not only from the basal tips but also from the lateral surface of the soma. The somatic dendrites run generally in a horizontal direction. Also the proximal stem of the apical dendrites gives off a fair number of side branches spreading out in the ganglionic layer. The more distal parts of the apical dendrite give off only a small number of side branches when piercing the supervening layers, and burst into terminal ramifications in the molecular layer. The axon descends and gives off collaterals in the usual manner (Ramón y Cajal, 1900b; von Economo and Koskinas, 1925; Balthasar, 1954; Gihr, 1968; Kaiserman-Abramof and Peters, 1972; Scheibel et al., 1974; H. Braak and E. Braak, 1976; Scheibel and Scheibel, 1978) (Fig. 2 E).

All of the aforementioned cell types still display a regular orientation of the cell body with reference to the cortical surface. Only a few cortical areas harbour modified pyramidal cells with dendrites almost equal in diametre and length which radiate into all directions.

Layer Pre-α of the entorhinal region, for example, accommodates such large multipolar nerve cells. Because of the characteristic pattern of dendritic arborization, layer Pre-α is often referred to as the layer of "stellate" cells (Ramón y Cajal, 1909; Lorente de Nó, 1934; Stephan, 1975). From our standpoint this designation evokes confusion. We classify these neurons with "modified pyramidal cells", since they always generate an axon from basal parts of their cell body which heads straight downwards to the white matter issuing on its way typical recurrent collaterals. The dendrites are richly studded with thorns. They emerge from the cell body with broad cone-shaped stems (Fig. 2 A). As the temporal proisocortex is approached,

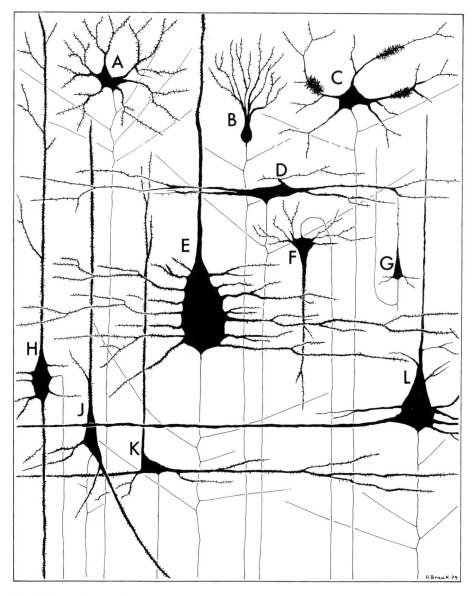

Fig. 2. Legend see p. 9

the entorhinal layer Pre-α sinks slowly into a deeper position and lies finally between the granular (IV) and the ganglionic layer (V) of the isocortex (see Chap. 5.3.2) (H. Braak, 1972a). Here, the gradual transformation of a multipolar nerve cell to a typical pyramidal cell can be observed. Apical and basal dendrites become better distinguishable the deeper the position of the Pre-α neuron. Finally, the nerve cells of Pre-α appear as well-formed large IIIc-pyramids (H. Braak et al., 1976).

The multipolar nerve cells of the fourth sector of the ammonshorn can serve as another example. Their dendrites also defy classification into apical and basal. Again the transition zone between the third and the fourth sector (see Chap. 5.1.2) shows the gradual appearance of a radially oriented main dendrite and a basal skirt. This and the complement of characteristic microdendrites (which is a common feature of both the typical pyramids of the third sector and the multipolar nerve cells of the fourth sector) allow one to classify the latter ones with modified pyramidal cells (Ramón y Cajal, 1893; Lorente de Nó, 1934; H. Braak, 1974a; Wenzel and Bogolepov, 1976) (Fig. 2 C).

Other types of modified pyramidal neurons show variations which are reminiscent of primitively organized pallial constituents of animals standing low in the phylogenetic scale. These forms show a bush-like spray of dendrites issuing out of the apical pole of the cell body. Basal dendrites (if present at all) appear underdeveloped. The axon arises from the base of the cell and pursues a straight course downwards (Ramón y Cajal, 1909; Poliakov, 1964/65; Northcutt, 1967; Capanna, 1969; Clairambault and Capanna, 1970; Kirsche et al., 1973; Kirsche, 1974; Ulinski, 1974; Turowski and Danner, 1977; Wallace et al., 1977). The granule cells of the fascia dentata as well as the extraverted neurons of other allocortical areas (Sanides, 1971; Sanides and Sanides, 1972) lack typical basal dendrites and belong to this group of modified pyramidal cells (Fig. 2 B).

◀ Fig. 2. Diagram showing an abnormally oriented pyramidal cell and various types of modified pyramidal cells. *A* Large multipolar nerve cell of the entorhinal layer Pre-α. *B* Granule cell of the fascia dentata with bushlike dendritic arbor. *C* Large multipolar nerve cell of the fourth sector of the cornu ammonis with characteristic microdendrites along circumscribed parts of the dendrites. *D* Solitary cell of Ramón y Cajal. This cell type with disk-like cell body is a characteristic component of layer IVb of the striate area. *E* Giant Betz pyramid with a great number of dendrites issuing from lateral parts of the cell body. These cells occur in layer Vb of the gigantoganglionic area. *F* Pyramid with normal dendritic arbor in abnormal orientation. *G* Small and slender pyramid of the fourth isocortical layer. After emergence of the first collateral the axon tapers off (short-axon pyramidal cell). *H–K* Pair of compass cells with two main dendrites disposed at different angles to each other. Cells of this type can mainly be encountered in the multiform layer. *L* Meynert pyramid with one unusually long basal dendrite. The dendrites are almost devoid of spines. Meynert pyramids occur in layer Vb of the striate area. Note the straight course of axons and collaterals of all the various types of modified pyramidal cells

Summing up, only a few traits are common to all these types of modified pyramidal cells. These features are the vertical orientation of the axon, its straight course, and the characteristic manner of its issuing collaterals. The dendrites furthermore appear generally decorated with spines. They usually originate from the cell body by way of a broad cone-shaped stem.

2.3 Non-Pyramidal Cells

Stellate cells prevail in the heterogeneous group of non-pyramidal cells. They show a great variability in shape and size (Ramón y Cajal, 1900a,b, 1902, 1903, 1909; Lorente de Nó, 1938; Peters, 1971, Poliakov, 1972/73; Jones, 1975; Feldman and Peters, 1978; Peters and Fairén, 1978; Ribak, 1978; Tömböl, 1978). Small stellate cells occur in all cortical layers; large ones can be encountered along the lower border of the third and the upper border of the fifth layer. Despite their variability the stellate cells have certain features in common. In general, they give off only a scant number of dendrites without spines which are already thin as they arise from the cell body. Bifurcation occurs often close to the soma. The dendrites run in all directions on a somewhat tortuous course and branch infrequently. They have a slightly varicose or beaded appearance (E. Braak, 1978b), a feature which is seemingly enhanced by suboptimal fixation (Williams et al., 1978). The dendrites do not become appreciably thinner as they stretch out distally. On rare occasions some isolated spiny appendages can be encountered along the dendritic processes of mature stellate cells; immature ones generally bear a greater number of such appendages (Meller et al., 1969; Le Vay, 1973; Lund, 1973; Jones, 1975; Lund et al., 1977). The axon emerges either directly from any point of the soma or from proximal parts of one of the dendrites whence it may run in various directions to ramify profusely in the vicinity of the parent soma.

In the Nissl preparation, the cell body of stellate cells shows a rounded outline (Peters and Kaiserman-Abramof, 1970). The ellipsoidal nucleus is usually in an excentric position. It contains a conspicuous nucleolus and often displays remarkable indentations of the otherwise indistinct nuclear membrane. Coarser Nissl granules are absent, the basophilic material is more or less homogeneously distributed throughout the peripheral parts of the soma.

As to the pigmentation of the stellate cells, two types are distinguishable; one is filled with coarse and intensely stained lipofuscin granules, the other completely lacks a pigmentation or contains only a few faintly tinged granules (Fig. 1 D, E). In the isocortex the majority of stellate cells are barren of pigment. Large stellate cells rich in pigment occur in the ganglionic and the multiform layer (V + VI). They are quite numerous in the proisocortex and rarer in isocortical "association" areas. The lower reaches of the corpuscular layer and subjacent parts of the pyramidal layer (II + III) accommodate a wealth of pigment-laden stellate cells which are of particularly small size. In contrast to the deep ones, they are preferentially found in the "association" areas of the isocortex. In places they attain such a packing density that they form a dark band in pigment preparations. This cell type has not yet been demonstrated in the cortex of subhuman primates. Considering the high concentration of these small pigment-laden stellate cells in the cortex which spreads over the opercula of the lateral cerebral sulcus, one is tempted to suggest that this cell type might be of particular significance for the higher functions of the human brain (H. Braak, 1974b, 1978a,b,c; E. Braak, 1976).

The number of stellate cells relative to that of the pyramids increases rapidly with phylogenetic advance and culminates in the refined isocortex of man. The stellate cells seem to develop late also in ontogeny and show a particularly prolonged postnatal differentiation (Mitra, 1955; Globus and Scheibel, 1967; Morest, 1969; Jacobson, 1969, 1970a,b, 1974, 1975; Purpura, 1975; Rakic, 1975; Rickmann et al., 1977).

The heterogenous group of non-pyramidal cells includes also two types of nerve cells which some people consider to be transitory. They can be found in greater numbers in the foetal pallium. Some disseminated cells of these types nevertheless persist and can constantly be encountered in the mature mammalian cortex.

The molecular layer contains the *horizontal cells of Cajal*. The dendrites of these cells spread out from the tips of the fusiform to polygonal cell bodies pursuing a course parallel to the cortical surface. The partly medullated axons — there may be more than one — run in the same direction and ramify within the molecular layer (Ramón y Cajal, 1891, 1909; Retzius, 1893, 1894; Oppermann, 1929; Noback and Purpura, 1961; Fox and Inman, 1966; Duckett and Pearse, 1968; Meller et al., 1968a; Sanides and Sas, 1970; Sas and Sanides, 1970; Marin-Padilla, 1972; Raedler and Sievers, 1975; Bradford et al., 1977; König et al., 1977; Rickmann et al., 1977). The nerve cells found in the molecular layer are almost devoid of lipofuscin accumulations.

The multiform layer accommodates another type of non-pyramidal cell: the *cells of Martinotti*. They display a pear-shaped to triangular cell body with dendrites disposed at various angles and an axon leaving the cell body at its uppermost tip. It ascends and divides into terminal ramifications in the molecular layer (Martinotti, 1890; Marin-Padilla, 1970a, 1971; Tömböl, 1972). The cells of Martinotti are barren of lipofuscin deposits.

3 The Three Standard Techniques Used in Architectonics

3.1 Preparations Stained for Nerve Cells (Nissl Preparations) as a Basis of Cytoarchitectonics

The lamination is most clearly understood in those regions collectively known as isocortex. This term refers to a cortex composed of six different laminae (Fig. 3). Delineation of the six laminae was first introduced by Berlin (1858). The details of the layers are subject to variations from area to area (Fig. 4). Avoiding the regional peculiarities, a scheme of a generalized isocortex can nevertheless be given to introduce the lamination pattern as seen in preparations stained for nerve cells (Brodmann, 1905/06, 1908, 1909; Vogt and Vogt, 1919).

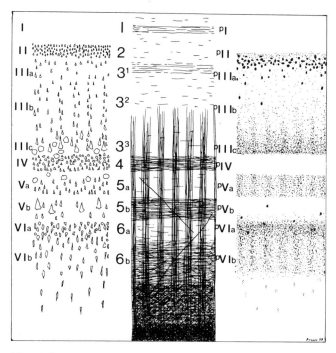

Fig. 3. Legend see p. 13

Fig. 3. Synopsis of the fundamental cyto-, myelo-, and pigmentoarchitectonic schemes. As concerns cyto- and myeloarchitectonics the nomenclature recommended by Vogt and Vogt (1919) is applied in our description. Isocortical layers as seen in Nissl preparations are marked with *Roman numerals*, those displayed by myelin preparations with *Arabic numerals*. The laminae in pigment preparations are characterized by a prefixed *"p"* and *Roman numerals*.

The cytoarchitectonic layers are termed from the pial surface inwards:

I	Molecular layer	*IV*	Granular layer
II	Corpuscular layer	*V*	Ganglionic layer
III	Pyramidal layer	*VI*	Multiform layer

Combinations such as the terms *magnopyramidal* or *magno(giganto)-ganglionic* point to a population of large to giant pyramidal cells in either the pyramidal layer or the ganglionic layer. The granular layer may be absent, patchy, narrow, of normal breadth, or particularly broad. The cortex is therefore denoted as being *agranular, dysgranular, intermediogranular, eugranular,* or *hypergranular*. Comparison between the breadth of sublayer IIIc and layer V gives the *externocrassior, equocrassus,* and *internocrassior* characteristic. The cortex may on the average be formed of small or large nerve cells (*parvocellular* or *magnocellular* type).

The myeloarchitectonic layers are termed from the surface inwards:

1	Zonal layer	*5a*	Intrastriate layer
2	Dysfibrous layer	*5b*	Internal stria
3	Suprastriate layer	*6*	Substrate and limiting layers
4	External stria		

The extension of radiate fibre bundles gives the *supraradiate* (up to 2) the *euradiate* (up to 3^2), and the *infraradiate* (up to 5) character.

The more or less clear delineation of the stripes of Baillarger add the *astriate, unistriate,* and *bistriate* characteristic. Fusion of both stripes leads to the *unitostriate* or *conjunctostriate* type. The *externostriate* cortex shows additionally the stripe of Kaes-Bechterew in the suprastriate layer. Comparison between the density of the external and internal stripe of Baillarger gives the *externodensior, equodensus,* and *internodensior* characteristic. The average myelin content of the cortex may be high or low (*typus dives* or *typus pauper*).

The pigmentoarchitectonic layers are termed from the surface inwards:

p_I	Molecular layer	p_{Va}	Ganglionic layer (outer part)
p_{II}	Corpuscular layer	$Ti(p_{Vb})$	Internal tenia
p_{III}	Pyramidal layer	p_{VI}	Multiform layer
$Te(p_{IV})$	External tenia		

The cortex may display both pallid teniae, or only the outer tenia, or it may be devoid of any light stripe. These pigmentoarchitectonic pictures are denoted as the *biteniate, uniteniate,* and *ateniate* character. The *stratiform* cortex shows in addition a third pallid stripe in the pyramidal layer. Comparison between the breadth of the external and internal tenia gives the *externoteniate, equoteniate,* and *internoteniate* characteristic. The average pigment content of the cortical nerve cells may be high or low (*typus obscurus* or *typus clarus*).

Structures running in parallel with the cortical surface are described as being narrow or broad, those perpendicularly aligned as being thin or thick

I. The *molecular layer* averages approximately one tenth of the total thickness of the cortex. It contains only a few nerve cells with perikarya different in size and shape. The layer is filled mainly by terminal ramifications of axons and dendrites (Jones and Powell, 1970b). The boundary towards the corpuscular layer is sharply drawn.

II. The *corpuscular layer* is generally a little less broad than the molecular layer. It is composed of tightly packed nerve cells of small size. The outline of the cell bodies as seen in Nissl preparations gives no sure criterion for the distinction of pyramidal cells from stellate cells. Small cortical nerve cells are therefore frequently referred to as "granule" cells. It is one of the shortcomings of Nissl preparations that small nerve cells cannot readily be classified with the group of either the pyramidal or the stellate cells. The boundary towards the pyramidal layer is blurred in some areas, but clearly recognizable in others.

III. The *pyramidal layer* makes up almost one third of the total thicknes of the cortex. It contains mainly well-formed pyramidal cells. A more or less abrupt increase in the average cell size allows one often to distinguish a sublayer IIIab (sublaminae parvo- et mesopyramidalis) from a sublayer IIIc (sublamina magnopyramidalis). Besides pyramidal cells there exist a fair number of stellate cells often with radially oriented spindle-shaped cell body (Jones and Powell, 1970c). Large to giant stellate cells occur close to the lower border of the layer.

IV. The *granular layer* is dominated by densely packed cells of small size. The breadth and the cell density of the layer are subject to pronounced regional variations. The layer can be absent or rather poorly sketched as for instance in the agranular motor cortex; it can also appear as a broad stripe as in the striate area (field 17, Brodmann).

V. The *ganglionic layer* occupies approximately one fifth of the total thickness of the cortex. It consists of medium-sized to large pyramidal cells intermingled with stellate cells. Close to its upper border particularly voluminous stellate cells can be encountered. The layer can often be subdivided into an upper cell-rich (Va) and a lower cell-sparse zone (Vb). The latter harbours unusually large pyramidal cells in a restricted number of cortical fields (Betz cells, Meynert cells).

VI. The *multiform layer* is of roughly the same thickness as the ganglionic layer. It is prevalently composed of modified pyramids with spindle-shaped or triangular cell bodies. The layer is often divisible into an outer denser (VIa, sublamina triangularis) and a lower looser zone (VIb, sublamina fusiformis) which gradually thins out towards the white substance.

The six laminae described can be found throughout most parts of the isocortex. These parts differing only in details from each other form the *homotypical* areas. *Heterotypical* areas, by contrast, reveal thoroughgoing variations of the normal layering or specializations of certain laminae. The heterotypical striate area (field 17, Brodmann) for instance shows laminae composed of highly characteristic types of nerve cells which do not occur in other parts of the brain and therefore cannot readily be homologized with isocortical laminae. A remarkably broad fourth layer densely filled with small nerve cells distinguishes the *hypergranular* cortex or *coniocortex*. A decrease in the normal number of layers can be observed in the heterotypical precentral motor cortex which is *agranular,* i.e., lacks the granular layer. *Dysgranular* and *intermedio-granular* fields with a rather thinly sketched just recognizable granular layer mediate between agranular areas on the one hand and *eugranular* ones on the other (Sanides, 1962) (Fig. 4).

Often it is advisable to compare the lower reaches of the pyramidal layer (IIIc) with the ganglionic layer (V). The overall impression of the tint of both laminae (dependent on the packing density, the size and the staining properties of the nerve cells) is assessed. In some cases IIIc is broader than V and well-filled with large pyramidal cells, elsewhere it is equally broad, and in still other areas it is inconspicuous as compared to a particularly broad ganglionic layer. These cytoarchitectural pictures are denoted as the *externocrassior,* the *equocrassus,* and the *internocrassior* characteristic (Sanides, 1962: *externopyramidal, equopyramidal,* and *internopyramidal* characteristic) (Fig. 4).

A *magnocellular* cortex is on the average formed of large nerve cells. This is opposed to a *parvocellular* one where "granules" prevail (Fig. 4). It should be remembered that the term "granule-cell" refers to only a small-sized cortical nerve cell which may be either a pyramidal cell or a stellate cell.

3.2 Preparations Stained for Myelin Sheaths as a Basis of Myeloarchitectonics

The cortex displays a variable content of myelinated fibres of various calibre and differently arranged. Meticulous analysis of myelin preparations therefore allows for a reliable cortical parcellation as well. Most of the laminae found in myelin preparations can be readily identified with corresponding layers of the Nissl preparations. Also the borderlines of cortical fields determined in myelin preparations generally coincide with those outlined in Nissl preparations. Hence both rather different methods supplement each other.

Unfortunately, only a few techniques can be used for myeloarchitectural analysis of the telencephalic cortex. These methods require frozen sections and regressive staining (see Chap. 8). Less sensitive methods developed for processing of paraffin material (Heidenhain-Woelcke; Loyez; Luxol-Fast-Blue) generally do not display fine intracortical fibres to advantage.

As regards numbering it is common practice to mark myeloarchitectonic laminae with Arabic numerals for clear distinction from cytoarchitectonic layers designated by Roman numerals (Fig. 3). The nomenclature used for myeloarchitectonic laminae is that of Vogt and Vogt (1919).

1. Subjacent a pallid oute zone (sublamina superficialis), the *zonal layer* contains a variable number of myelinated fibres running parallel with the cortical surface. Well-stained preparations display distinguishing characteristics of the first layer in different areas. In routine preparations, unfortunately, the variations in the fibre content are often not displayed to advantage. The myeloarchitectonic traits of the first layer are therefore usually not taken into account for delineation of cortical areas. Primary sensory and motor fields are particularly rich in tangential fibres whereas "association" areas are generally sparsely endowed with such a plexus.

2. The *dysfibrous layer* impresses itself as a pallid stripe almost devoid of myelinated fibres.

3. The *suprastriate layer* shows fibres which preferentially pursue a course parallel to the cortical surface. Their numbers increase considerably as one descends through the layer. Often this provides a basis for an arbitrary subdivision of the layer into 3^1, 3^2, and 3^3. In places, the upper reaches of the layer contain a concentration of fibres which constitute the *stripe of Kaes-Bechterew* (von Bechterew, 1891; Kaes, 1907). It can be encountered in sensory core fields or their surrounding areas. The dysfibrous and the suprastriate layer in particular show a prolonged period of maturation (Haug et al., 1976). The myelin content of the superficial layers may increase up to the 25th year (Kaes, 1907).

Bundles of more or less tightly aggregated fibres which are arranged perpendicular to the surface generally go as far as the boundary between 3^2 and 3^3 (radiate bundles).

4. The *external stria* — the *outer stripe of Baillarger* — is generally an impressive clearly delimited band of densely packed fibres.

5a. The *intrastriate layer* is in most isocortical areas relatively poor in horizontal fibres, thereby contrasting with the two bordering lines of Baillarger.

5b. The *internal stria* — the *inner stripe of Baillarger* — is again a dense plexus of horizontally oriented fibres.

6. The sixth layer can be further subdivided by starting out with the relatively pallid *substriate lamina* (6aα); this is followed by the *external*

and *internal lamina limitans* (6aβ and 6bα), each showing increasing wealth of horizontal fibres thereby merging with the white substance (6bβ).

Variations in the arrangement and number of the radial and tangential fibres define the limits of cortical areas. In the mature isocortex the bundles of radial fibres generally do not surpass the boundary between 3^2 and 3^3, a pattern which is referred to as the *euradiate* characteristic. Proisocortical areas (see Chap. 6), by contrast, show either a *supraradiate* pattern with bundles extending as far as the first layer, or an *infraradiate* arrangement, where bundles cannot be followed beyond the limits of the inner stripe of Baillarger (Fig. 4).

Two well-developed bands of Baillarger, each framed by relatively fibre-sparse zones, represent the *bistriate* type of cortex. The breadth and the fibre content of both bands vary considerably. A denser outer stripe, equal myelin density of both lines, or accentuation of the inner stripe is noted respectively as typus *externodensior, equodensus,* or *internodensior* (Fig. 4).

Unistriate fields display only the outer stripe of Baillarger. The inner one is present but cannot be outlined because of the high myelin density of the substrate lamina. Considerable increase in the fibre content can also occur in the intrastriate layer; this is consequently linked with a more or less pronounced veiling of both stripes of Baillarger. Provided that the substrate lamina is also filled with myelinated fibres this pattern is denoted as the *astriate* characteristic. *Extremostriate* fields are defined by the appearance of the stripe of Kaes-Bechterew in addition to the bands of Baillarger. *Unitostriate* areas display a broad medullated band bounded above and below by pallid zones. The band consists of both Baillargers and an intrastriate layer in between which is rich in relatively "solitary" fibres. *Conjunctostriate* areas show the same pattern but with an intrastriate layer densely filled with thin "ground" fibres (Fig. 4). Areas devoid of an inner stripe of Baillarger represent the rarely occurring *singulostriate* type.

Particularly rich endowment of myelinated fibres — horizontal as well as radiate — is referred to as *typus dives,* and is opposed to the *typus pauper* with its usually fibre-sparse cortex (Fig. 4). In the evolutionary elaboration of the telencephalic cortex there is in general a remarkable increase in the average myelin content (Bishop and Smith, 1964; Sanides, 1970, 1972).

3.3 Preparations Stained for Lipofuscin Granules as a Basis of Pigmentoarchitectonics

Architectonic studies of large brains such as that of the human being are greatly facilitated by using a selective staining technique for lipofuscin pigments (H. Braak, 1978a,e). Because of the fact that only a single cytoplasmic component is stained the thickness of the sections can be considerably enlarged (800 μm) thereby facilitating the processing and the examination of an extensive material. Properly processed preparations leave the pigment in glial and endothelial cells unstained. Only nerve cells are marked by their lipofuscin deposits. The great number of superimposed cortical neurons which in this way can be scrutinized at a glance allows one to delineate with ease the borderlines of both cellular layers and cortical areas. Due to the remarkable thickness of the brain slices the preparations are particularly well-adapted for low-power examination with the stereomicroscope.

Different types of cortical neurons generally show a highly characteristic pattern of pigmentation (Obersteiner, 1903, 1904; Vogt and Vogt, 1942; Mannen, 1955; Creswell et al., 1964; H. Braak, 1971, 1978a). The numerical increase of intracellular lipofuscin granules which regularly and gradually occurs as age advances does not change the pattern of pigmentation.

Pigment preparations show an easily recognizable series of laminae which are characterized by a prefixed "P" and Roman numerals (Fig. 3).

P_I. The *molecular layer* can be subdivided into three zones of different width. Subjacent to the narrow and pallid external glial layer (see Chap. 7.1.1) there appears a broad zone containing the cell bodies of the modified astrocytes. The pigment stored in these astrocytes is modestly tinged by aldehydefuchsin — in contrast to the astrocytic pigment found elsewhere. Only a sparse number of astrocytes and occasionally some horizontal cells of Cajal almost devoid of pigment can be encountered in the lower half of the molecular layer. It is therefore a light zone.

P_{II}. The *corpuscular layer* is composed of tiny pyramids and small stellate cells. Most of the sparsely pigmented pyramids lie within the upper parts of the layer. Both types of stellate cells occur. Those almost devoid of pigment inclusions can only be shown by counterstaining with a basic dye. The pigment-laden variety fills mainly the lower half of the layer as well as subjacent parts of the pyramidal layer.

P_{III}. The *pyramidal layer* accommodates well-formed pyramids increasing in size and pigmentation from the upper border to the lower one. Often a more or less sudden break in pigmentation allows one to distinguish a lighter upper (P_{III}ab) from a lower darker part (P_{III}c). The small pyramidal cells within the upper reaches of the third layer appear late in ontogeny and offer many traits characteristic of particularly elaborate pyramids (E. Braak et al., 1980). In the senescent human brain these nerve cells are almost selectively afflicted by a striking pathological change in that the proximal axon develops a spindle-shaped enlargement tightly filled with lipofuscin granules (H. Braak, 1979a, E. Braak et al., 1980). The densely pigmented lower reaches of the layer show a radially oriented weak striation which is caused by bundles of myelinated fibres piercing it.

Pigment-laden IIIc-Pyramids. In addition to the common variety of large third-layer pyramids with weakly tinged and evenly distributed pigment accumulations there occurs in a few isocortical areas a conspicuous population of large IIIc-pyramids which accumulate lipofuscin granules in rounded aggregates. The axon hillock and large parts of the cytoplasm remain free of pigment. These cells, which catch the eye in pigment preparations, keep relatively large distances between each other. The position they prefer is the lower border of IIIc, but they can be found also in great number within the pallid subjacent zone (Te).

Te. The *external tenia*, a pallid stripe in pigment preparations, is filled with sparsely pigmented tiny pyramids and stellate cells devoid of pigment. The outer tenia can be present also in cortical areas lacking a granular layer (IV). In these cases the pigment-barren stripe generally covers the lower reaches of the pyramidal layer formed of large IIIc-pyramids devoid of pigment. In general, the position of the light stripe corresponds to that of the outer band of Baillarger, but there are exceptions to this rule.

P_V. The *ganglionic layer* is generally split into a feebly tinged P_{Va} rich in cells, and a pallid and cell-sparse P_{Vb} — each approximately equally wide. Densely packed pyramids with a pattern of pigmentation similar to that of the common type of IIIc-pyramids prevail in P_{Va}. The layer displays a clear radially aligned striation resulting from bundles of myelinated fibres in between columns of pyramidal cells.

Ti. The lower half of the ganglionic layer appears again as a light stripe which consequently is denominated the *internal tenia*. Its position corresponds generally to that of the inner line of Baillarger. The layer is populated by only a modest number of weakly pigmented pyramids and a few large and pigment-rich stellate cells dotted about here and there. This type of stellate cell, which also occurs in the multiform layer, shows a marked numerical increase as one approaches the older parts of the telencephalic cortex.

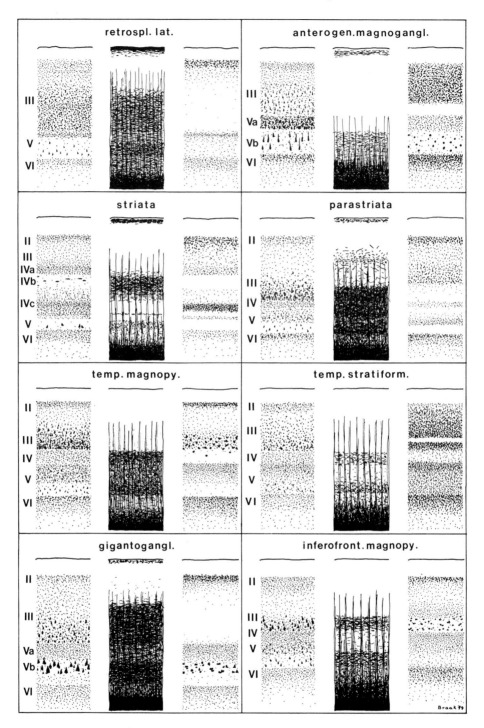

Fig. 4. Legend see p. 21

◄ **Fig. 4.** Diagrammatic representation of the appearance of eight areas of the human telencephalic cortex in the Nissl, the myelin, and the pigment preparation. The characteristics of the areas can be found in the following list which may serve for an introduction into the nomenclature needed for the description of architectural entities

Area retrosplenialis lateralis (see Chap. 6.1)			Area anterogenualis magnoganglionaris (see Chap. 6.2)		
parvocellular	dives	clarus	magnocellular	pauper	obscurus
agranular	propeastriate	biteniate	agranular	only int. stria	biteniate
externocrassior	externodensior	externoteniate	internocrassior	internodensior	internoteniate
	supraradiate			infraradiate	

Area striata (see Chap. 7.1.1)			Area parastriata (see Chap. 7.1.2)		
parvocellular	dives	clarus		dives	
hypergranular	singulostriate (?)	biteniate	eugranular	subconjuctostr.	biteniate
externocrassior	externodensior	externoteniate	externocrassior	extremostriate	externoteniate
	euradiate			euradiate	

Area temporal. magnopyr. centr. (see Chap. 7.2.4)			Area temporalis stratiformis (see Chap. 7.2.5)		
				pauper	obscurus
eugranular	bistriate	biteniate	eugranular	bistriate	uniteniate
equocrassus	equodensus	equoteniate	equocrassus	equodensus	stratiform
	euradiate	magnopyramidal		euradiate	

Area gigantoganglionaris (see Chap. 7.4.1)			Area inferofront. magnopyr. centr. (see Chap. 7.4.3.1)		
magnocellular	dives	clarus			
agranular	astriate	biteniate	intermediogran.	propeunistriate	biteniate
internocrassior		externoteniate	externocrassior	internodensior	equoteniate
	euradiate	giganto-ganglionic		euradiate	magno-pyramidal

These frequently used terms for cortical characteristics mean:

agranular	absence of layer IV (granular layer)
astriate	both lines of Baillarger cannot be outlined
bistriate	two lines of Baillarger recognizable
clarus	light, sparsely endowed with pigment
conjunctostriate	fusion of the lines of Baillarger
dives	rich, high myelin content
equocrassus	layers IIIc and V equally broad
equodensus	both lines of Baillarger equally dense
equoteniate	both teniae equally broad
eugranular	well-developed layer IV (granular layer)
euradiate	radiate bundles extend to the upper border of lamina 3^3
externocrassior	layer IIIc is broader than layer V
externodensior	the outer line of Baillarger is denser than the inner one
externoteniate	the outer tenia is broader than the inner one
extremostriate	the stripe of Kaes-Bechterew is recognizable
gigantoganglionic	giant pyramidal cells in layer Vb
hypergranular	particularly broad layer IV (granular layer)
infraradiate	radiate bundles extend to the upper border of lamina 5b
intermediogranular	tenuous layer IV (granular layer)

Legend to Fig. 4 continued

internocrassior	layer V is broader than layer IIIc
internodensior	the inner line of Baillarger is denser than the outer one
internoteniate	the inner tenia is broader than the outer one
magnocellular	large cells prevail
magnopyramidal	large pyramidal cells in layer IIIc
obscurus	dark, richly endowed with pigment
parvocellular	small cells prevail
pauper	poor, low myelin content
prope	almost
singulostriate	with only the outer line of Baillarger
stratiform	with a pallid stripe in lamina ^{P}IIIc
supraradiate	radiate bundles extend to lamina 1−2
unistriate	the outer line of Baillarger recognizable, high myelin content of laminae 5b and 6
uniteniate	only the outer tenia recognizable

Pigment-laden Vb-Pyramids. In places, sublayer ^{P}Vb can be found more or less richly stocked with unusually large pyramids which accumulate vast amounts of pigment. These cells represent the Betz pyramids (Betz, 1874, 1881). Their lipofuscin granules assemble to form tightly packed agglomerations.

The axon hillock is devoid of pigment. The size and shape of the Betz cells vary pronouncedly (Campbell, 1905; Lassek, 1940; Bailey and von Bonin, 1951; Solcher, 1958; Gihr, 1968; H. Braak and E. Braak, 1976). The Betz cells have therefore often been considered to be only a variant of fifth-layer pyramids, as their appearance in Nissl preparations does not provide any basis for their unequivocal distinction from other large pyramids (Walshe, 1942; Kaiserman-Abramof and Peters, 1972). In pigment preparations, nevertheless, Betz cells appear as a homogeneous class of pyramids which, regardless of their size, can readily be separated from other pyramids by their characteristic pigmentation (H. Braak, 1976a; H. Braak and E. Braak, 1976).

P_{VI}. The *multiform layer* is dominated by modified pyramidal cells with varying amount of pigment. It appears often as a bipartite layer with a darker upper and a lighter lower zone, the latter frequently merging with the white substance.

Pigment preparations resemble to a certain degree Nissl preparations since they display layers of nerve cells. They also bear a certain likeness to the negative image of myelin preparations because of the existence of the light teniae, the position of which often corresponds to that of the lines of Baillarger. The two teniae show, like the Baillarger stripes, a considerable variability in breadth and in the formation of their borderlines.

Biteniate areas clearly display both teniae. A broader outer tenia, equal breadth of both lines, and a broader inner tenia are denoted respectively as an *externoteniate, equoteniate,* or *internoteniate* characteristic. The upper border of the outer tenia often appears particularly blurred in the extremely externoteniate areas (Fig. 4).

Uniteniate fields are characterized by the existence of only the outer tenia, whereas *ateniate* areas lack any pallid band. *Propebiteniate* fields display a distinct external tenia and a blurred internal one. *Propeuniteniate* areas show an attenuated and hazy outer band and only an ill-defined remnant of the inner one. In the *propeateniate* areas finally only a poorly represented and almost imperceptible external tenia can be found (Fig. 4).

Stratiform areas exhibit a narrow and pallid band running in parallel with the cortical surface. In general, the line is located within the limits of the lower third of the pyramidal layer (P_{III}).

The *typus obscurus* characterizes a cortex, the cellular laminae of which are filled with richly pigmented pyramids. The *typus clarus* on the contrary marks a pallid cortex with pyramids sparsely endowed with pigment (Fig. 4).

4 The Main Subdivisions of the Telencephalic Cortex

The cortex can be divided into allocortical and isocortical territories (Vogt and Vogt, 1919). In man isocortical areas cover the major parts of the telencephalon. Ontogenetically they show an almost uniform development and share therefore general principles of structure.

The allocortical areas which comprise the remaining parts of the cortex display rather marked differences of structure. Allocortical areas range from simply organized fields to highly refined ones with a greater number of laminae than is commonly seen in isocortical areas. The partly extended stretch of cortex mediating between the allocortex and the isocortex is often difficult to classify. Most authors agree in delineating two borderline zones. The periallocortex accompanies the allocortex sensu stricto (Filimonoff, 1947). It is followed by the proisocortex, which is more closely allied to the isocortex (Vogt and Vogt, 1956; Sanides, 1962, 1963, 1964; Stephan, 1963, 1975; Stephan and Andy, 1970; Kirsche, 1972).

Because of its heterogeneous nature the allocortex sensu stricto defies clear definition (Stephan, 1975). The changing features of the allocortex clearly influence the architecture of the expanded stretch of cortex adjacent to it (Kirsche, 1972). As one passes from the allocortex to the isocortex, the characteristics of the allocortex generally disappear in definite stages, whereas those of the isocortex become gradually more distinct. Most of the allocortical laminae are sufficiently marked by a characteristic pigmentation to allow for their unequivocal distinction from isocortical layers. In general, the border of the allocortex can therefore be clearly outlined. We define the allocortex sensu stricto as being totally composed of allocortical laminae. The number of layers varies considerably from one area to another. A great number of allocortical fields are also marked by a thick plexus of myelinated fibres in the molecular layer − a feature which often allows one to define them macroscopically (Meynert, 1868, 1872).

As a rule, some of the allocortical layers transgress the limits of their parent territories to interdigitate with a set of isocortical layers. The complex stepwise change in lamination pattern thereby established ranges from marginal fields with only a single isocortical layer to areas with only a single allocortical one. These changes in layering allow one to trace decisively the borderline of the periallocortex, which is here defined as an admixture of both allocortical and isocortical laminae (H. Braak, 1978c).

The adjoining isocortex may still show for a variable distance some features which can be encountered only in these borderline zones, such as a band-like appearance of the ganglionic layer in the Nissl preparation or only a small number of second-layer pigment-laden stellate cells in the pigment preparation. This marginal zone is designated the proisocortex. Isocortical areas generally display an euradiate characteristic whereas proisocortical fields may also show the infraradiate or supraradiate type. But unfortunately, there is no clear criterion for defining the boundary of the proisocortex with the isocortex sensu stricto.

The total transitional zone including the periallocortex and the proisocortex is often referred to as the mesocortex (M. Rose, 1927a,b; Brockhaus, 1940; Kirsche: mesoneocortex, 1974; Stephan, 1975).

Allocortex sensu lato ⎰ Allocortex sensu stricto
⎱ Periallocortex ⎫
⎰ Proisocortex ⎭ Mesocortex
Isocortex sensu lato ⎱ Isocortex sensu stricto

5 The Allocortex

5.1 The Hippocampal Formation

One of the main constituents of the allocortex is the hippocampal formation. In the human brain it is particularly well-developed in the temporal lobe and occupies the floor of the inferior horn of the lateral ventricle from about the level of the corpus amygdaloideum up to the splenium of the corpus callosum where it extends on to the ridge with a rudimentary supracallosal portion.

Structural differences betoken its organization into three major parts: the fascia dentata, the cornu ammonis, and the subiculum (Ramón y Cajal, 1893, 1909; Lorente de Nó, 1934; Vogt and Vogt, 1937; Rose, 1938; H. Braak, 1972b, 1974a; Chronister and White, 1975; Stephan, 1975).

5.1.1 The Fascia Dentata

The fascia dentata accompanies the ammonshorn as a small frequently indented gyrus, bordered by the hippocampal and the fimbriodentate sulcus. Anteriorly, it spreads over the free surface of the uncus constituting the limbus Giacomini. Its posterior extremity forms the fasciola cinerea which can be followed up to the splenium corporis callosi (Klingler, 1948).

The fascia dentata consists of three laminae: the molecular layer, the granular layer, and an ill-defined plexiform layer.

Small nerve cells usually referred to as the "granule" cells dominate in the fascia dentata. Here in contrast to its meaning in other parts of the telencephalic cortex (where it only refers to the particular smallness of neurons) the term designates a well-defined type of cortical nerve cells. The perikarya of the granule cells assemble tightly together and form a clear-cut band (Fig. 5). Most of the cell body is occupied by a clear rounded

---

Fig. 5. Fascia dentata (*Fd*) and cornu ammonis (*CA 1–4*) of man. *Upper half* Nissl preparation (15 μm). *Lower half* Pigment preparation (800 μm). Note the clear borderlines between the various sectors of the cornu ammonis in the pigment preparation. Also the stratum oriens filled with pigment-laden basket cells shows up clearly

The Fascia Dentata

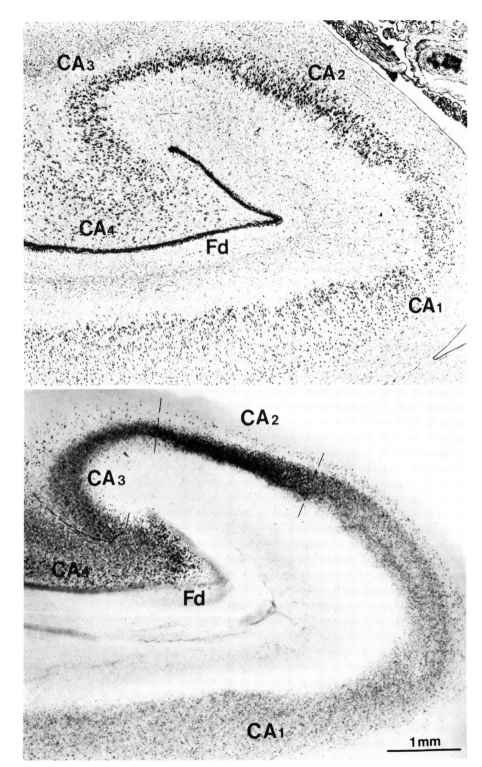

Fig. 5. Legend see p. 26

nucleus. The perikaryal cytoplasm is confined to a narrow rim which is poor in basophilic material and contains only a few feebly tinged pigment granules. A few dendrites emerge from the apical pole of the pear-shaped or ovoid cell bodies. They extend towards the surface, bifurcating repeatedly at a short distance from the soma (Frimmel et al., 1975; Lindsay and Scheibel, 1976). The dendrites are richly decorated with spiny appendages. The basal tip of the soma shows a distinct axon hillock from which a relatively thick axon is generated which enters the cornu ammonis and forms synaptic contacts with the pyramids of the third and the fourth sector (mossy fibre system: Sala, 1891; Schaffer, 1892; Ramón y Cajal, 1893, 1909; Lorente de Nó, 1934; Maske, 1955; Fleischhauer and Horstmann, 1957; Fleischhauer, 1958, 1959; Timm, 1958; Blackstad and Kjaerheim, 1961; Hamlyn, 1962; Friede, 1966; Otsuka and Kawamoto, 1966; Haug, 1967; Blackstad et al., 1970; Haug et al., 1971; Lynch et al., 1973; Wenzel and Bogolepov, 1976; Amaral, 1978; Gaarskjaer, 1978).

A small number of stellate cells with smoothly contoured dendrites and local axon is scattered throughout the granular layer and its immediate vicinity. The plexiform layer contains the majority of stellate cells, the axons of which ramify within the granular layer. These stellate cells are either crammed with lipofuscin granules or are almost devoid of pigment. The plexiform layer cannot be clearly distinguished from the fourth sector of the ammonshorn (Ramón y Cajal, 1909; Laatsch and Cowan, 1966; Stephan, 1975; Amaral, 1978).

The general smallness of the nerve cells allows one to classify the fascia dentata with the parvocellular cortices. It possibly maintains also a functional relationship to coniocortices in other parts of the brain (Hassler, 1967).

5.1.2 The Cornu Ammonis

The extent of the cornu ammonis increases greatly from the gyrus fasciolaris to the uncus.

On cross-sections, it displays three to four laminae: the molecular layer, the pyramidal layer (which in places can be split into an outer and an inner lamina) and the stratum oriens. The last layer borders upon the white matter which in the ammonshorn forms the alveus (Burdach, 1819–26).

The most characteristic cell constituents are large pyramidal cells which lie closely together. Their somata contain diffusely distributed basophilic material and only a modest number of lipofuscin granules. These are preferably arranged in several rows parallel to the lateral surfaces. The small and only feebly tinged granules do not tend to amalgamate (H. Braak, 1974a).

From the apical pole of the cell body arises a short stem which arborizes richly. The opposite pole gives rise to a great number of basal dendrites which extend into the stratum oriens. They form an extensive dendritic domain since their trunks are for a short distance more or less horizontally oriented. The human ammonshorn pyramids differ in this respect from the slender "double-bush" cells of small rodents with bushels of radially oriented apical and basal dendrites (Ramón y Cajal, 1909; Niklowitz and Bak, 1965; Wenzel et al., 1972, 1973, 1977; Englisch et al., 1974; Frotscher et al., 1975, 1978a,b; Kunz et al., 1976; Minkwitz, 1976a). Both the apical and the basal dendrites are studded with thorns. The axon is fairly thick and gives off a number of collaterals before entering the alveus. The collaterals preferably terminate in the stratum oriens.

The main constituents of the stratum oriens are the basket cells with large spindle-shaped cell bodies. The Nissl substance is diffusely dispersed and mostly restricted to the marginal parts of the soma. A great number of cells in the stratum oriens are crammed with intensely stained lipofuscin granules (H. Braak, 1974a). The basket cells issue wide spreading dendrites from opposite poles of the cell body. The dendrites are generally smoothly contoured. Only on rare occasions can some isolated spines be found. The axon ascends and splits up into its terminal ramifications between the cell bodies of the pyramidal cells (Sala, 1891; Ramón y Cajal, 1909; Minkwitz, 1976b).

The subdivision of the ammonshorn into four sectors as proposed by Lorente de Nó (1934) using the classical neurohistological techniques is confirmed by pigmentoarchitectonics (Fig. 5).

The *fourth sector* (CA4) fills up the hilus of the fascia dentata and does not show layering. Its dominating constituents — modified pyramidal cells — do not show a regular arrangement and orientation. The various dendrites have about the same diametre and length. Limited parts of them are densely studded with microdendrites (Wenzel and Bogolepov, 1976).

The stratum oriens vanishes gradually within CA4. The basket cells are shifted through the modified pyramids towards the plexiform layer of the fascia dentata. Hence within the marginal parts of CA4, local circuit neurons of both the ammonshorn and the fascia dentata come together to form the aforementioned ill-defined stripe which is part of the fascia dentata and the ammonshorn as well (Stephan, 1975).

The *third sector* (CA3) shows a unifold layer of well-formed and regularly oriented pyramids. The hallmark of these cells is long microdendrites which cover confined portions of both the apical and the basal dendrites. The microdendrites are sites of contact with the mossy fibres. As a result of the regular arrangement of the CA3-pyramids the molecular layer can be further divided into a substratum radiatum (Meynert, 1868) harbouring terminal ramifications of the mossy fibre system, a substratum lacunosum formed of the finer dendritic twigs which are studded with spines of the common type, and a substratum eumoleculare (Stephan, 1975) enclosing the endbranches of the apical dendrites. In pigment preparations the pathway of the mossy fibres is indicated by astrocytes filled with lipofuscin granules which are stainable by aldehydefuchsin (H. Braak, 1974a).

The *second sector* (CA2) displays already a splitting of the pyramidal layer into a superficial and a profound lamina, each composed of several rows of pyramidal cells. The profound pyramids are loosely distributed and contain a modest number of lipofuscin granules whereas the superficial ones are densely placed and richly endowed with pigment. The dark band of the superficial pyramids in particular permits easy delineation of CA2 in pigment preparations (Fig. 5). In Nissl and myelin preparations by contrast CA2 is not displayed to advantage and is difficult to delineate.

The extent of the first sector (CA1: Sommer's sector) increases markedly with phylogenetic advance (Stephan, 1975) In the human brain it is a particularly ample part of the ammonshorn. The pyramidal cells are arranged in two layers, stratum profundum and stratum superficiale, both being formed of several rows of sparsely pigmented pyramids which keep relatively far apart from each other (Lorente de Nó, 1934; Vogt and Vogt, 1937; H. Braak, 1974a). The stratum oriens seems unchanged.

5.1.3 The Subiculum

The subiculum (designation introduced by Burdach, 1819) forms the third part of the hippocampal formation. It is located between the ammonshorn on the one side and the presubiculum on the other. Also the subiculum undergoes considerable elaboration with phylogenetic advance (Stephan, 1975). At least in the primate brain, the subiculum appears to be the only source of efferent fibres from the hippocampal formation to other parts of the telencephalic cortex and contributes to subcortical fornix fibres as well (Rosene and van Hoesen, 1977; Schwerdtfeger, 1979) (Fig. 12).

Broadest at its anterior extremity where it spreads over a large part of the uncinate gyrus the subiculum narrows down continually in the posterior direction. After reaching the ridge of the corpus callosum, it consists of merely a small band of cells accompanying the remnant of the ammonshorn.

The *molecular layer* is remarkably broad. Its upper reaches are particularly rich in myelinated fibres. A substratum radiatum is lacking, a feature which permits tracing the borderline between CA1 and the subiculum in preparations stained for nerve cells.

The following band of cells is broader than in other parts of the hippocampus. The cellular layers enclose two highly characteristic laminae which only occur within the limits of the subiculum. Large pyramidal cells prevail in both of these laminae.

The *external pyramidal layer* appears as a particularly broad band formed of several rows of pyramidal cells which keep large distances between each other. The perikarya contain finely dispersed Nissl substance and a few disseminated lipofuscin granules. The axon and the basal dendrites are barren of pigment. The stout apical dendrite by contrast displays a conspicuous agglomeration of pigment which is spindle-shaped and fills up a circumscribed portion of the cellular process close to the soma (H. Braak, 1972b) (Fig. 6).

Pigment accumulations within dendrites can only rarely be encountered within the central nervous system. As an example, the large multipolar nerve cells of the nucleus alaris in the dorsal glossopharyngeus and vagus area exhibit similar dendritic pigment spindles (H. Braak, 1972c). Within the telencephalic cortex the subiculum is the only region which is distinguished by this feature. It can therefore easily be outlined in pigment preparations. The reason for the development of dendritic pigment is unknown at present.

The *internal pyramidal layer* is dominated by medium-sized pyramids. As compared to the external ones the longer axes of their cell bodies are less regularly oriented. The perikarya contain a great number of tightly packed lipofuscin granules which form a bowl-shaped mass close to the nucleus. These cells are devoid of dendritic pigment spindles (Fig. 6).

Large stellate cells with ovoid or spindle-shaped cell bodies are disseminated throughout both the external and the internal pyramidal layer. Most of them are almost barren of pigment; only a few store a greater number of lipofuscin granules.

In sum, the subiculum appears clearly dominated by large and highly specialized pyramidal cells. It is therefore classified with the magnocellular cortices. In comparison to CA1, it appears even more elaborate and is therefore denoted as the refined magnocellular core of the hippocampus.

Besides the endogeneous cellular layers which stamp the subiculum a variable number of foreign laminae penetrate into the region from the surrounding fields. This results in an often intricate lamination pattern. The thickness of both the endogeneous and the foreign laminae changes continually, a fact which accounts for incessant local variations. The parcellation of the subiculum is therefore based on only the basic changes in cortical lamination. Five major areas can be outlined.

In *area subicularis lateralis centralis* (s_{1c}, Fig. 12) a band of outer CA1-pyramids is squeezed between the molecular layer and the endogeneous

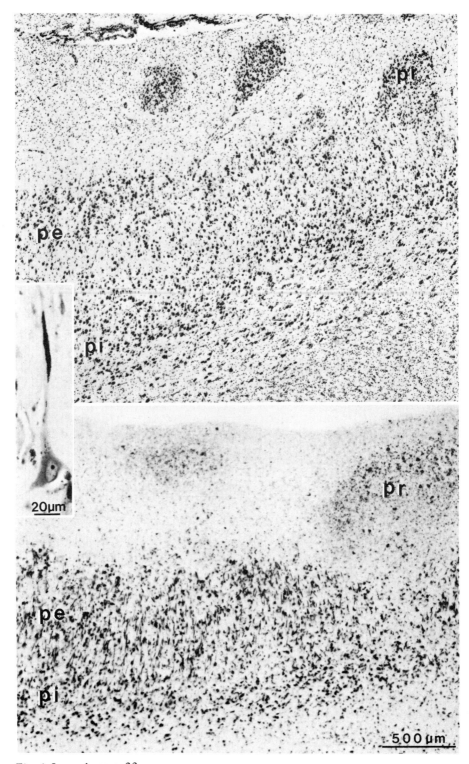

Fig. 6. Legend see p. 33

pyramidal layers of the subiculum. As one goes from the lateral to the medial border of the area, the layers of subicular pyramids gradually increase in breadth at the expense of CA1. Frontal sections therefore show two wedge-shaped formations separated by a borderline which runs obliquely through the cortex.

In *area subicularis lateralis marginalis* (s_{lm}, Fig. 12) the outer subicular pyramids have vanished so that the cortex is composed of only the molecular layer, the CA1-pyramids and the deep subicular pyramids.

In *area subicularis medialis oralis* (s_{mo}, Figs. 7, 12) irregularly spaced clouds of presubicular cells crop up within the broad molecular layer. Size and shape of these islands vary considerably. Bell-shaped clouds are frequently found with their longer axes oriented perpendicular to the cortical surface. Subjacent to the two distinct layers of subicular pyramids there is a narrow zone relatively empty of nerve cells and another cellular layer, the constituents of which belong to a deep entorhinal layer (Pri-γ). Accordingly, the layer appears as a dense and cell-rich stripe at the medial border of the field but becomes tenuous as one goes laterally.

As compared to the entorhinal region the subiculum is particularly extended in the posterior direction. Accordingly, the *area subicularis medialis caudalis* (s_{mc}, Fig. 12) displays the same lamination pattern as the foregoing field except for the deep entorhinal component. The extended territory where islands of presubicular nerve cells transgress the limits of the subiculum can be classified with either the presubiculum (von Economo and Koskinas, 1925; Stephan, 1975) or the subiculum (Ramón y Cajal, 1903, 1909; Rose, 1935). As the endogeneous laminae of the subiculum form the dominating components of the cortex, the territory is here designated as part of the subiculum (H. Braak, 1972b, 1978d).

The clouds of small presubicular nerve cells do not extend on the ridge of the corpus callosum. Close to the posterior extremity of the presubiculum the *area subicularis posteropolaris* (s_{pp}, Fig. 12) appears where clouds of presubicular cells mix with scattered clusters of CA1-pyramids. The deep layers again are formed of characteristic outer and inner subicular pyramidal cells.

◄ **Fig. 6.** Subiculum of man close to its medial margin. *Upper half* Nissl preparation (15 μm). *Lower half* Pigment preparation (200 μm). The extremely broad molecular layer contains some clouds of the parvopyramidal presubicular layer (*pr*). The cellular laminae of the subiculum can be divided into an external and an internal pyramidal layer (*pe* and *pi*). The pigment preparation reveals a distinguishing characteristic of the external subicular pyramidal cells which contain a spindle-shaped pigment agglomeration within circumscribed parts of the apical dendrites (*inset at left-hand margin*)

5.2 The Presubiculum

The presubiculum is a small and elongated region which borders laterally on the subiculum and medially on either the entorhinal region or the temporal and occipital proisocortex. Anteriorly the region covers a small part of the uncinate gyrus. Close to its posterior extremity, it splits in two: one part follows the subiculum and vanishes close to the splenium corporis callosi; the other portion is wedged between the retrosplenial region on the one side and the occipital properistriata on the other.

The presubicular cortex consists of only one endogeneous layer formed of particularly small nerve cells. The remaining laminae belong to the allocortex or isocortex outside the region. Similar to the formation of the subicular fields, the foreign layers transgress the limits of the presubiculum to a variable degree. This results in a characteristic lamination pattern which allows for sure delineation of presubicular fields (H. Braak, 1978d).

The characteristic small nerve cells of the presubiculum form in the posteromedial parts of the region a solid layer but anterolaterally tend to disintegrate into clouds of different size and shape (Fig. 7). This tendency is particularly expressed in the human brain and only mildly existent in those of subhuman primates (Altschul, 1933). The tiny nerve cells of the parvopyramidal presubicular layer do not always display the typical features of pyramids and can be better classified with the group of modified pyramidal cells (Ramón y Cajal, 1903, 1909; Lorente de Nó, 1934). The polygonal cell bodies of these neurons contain relatively large and pallid nuclei, finely dispersed basophilic material and a few coarse pigment granules. These are composed of dense parts and light vacuoles which can readily be recognized with the light microscope. Complex pigment granules of this type do not occur in the nerve cells of neighbouring areas, thereby allowing sure identification of even widely disseminated and displaced cell bodies of the parvopyramidal layer. This variety of pigment can regularly be encountered in lamina $IVc\beta$ of the striate area which is also formed of tiny nerve cells and possibly also functionally related to the parvopyramidal presubicular layer.

Besides the prevailing pyramids there exists a sparse population of medium-sized stellate cells which are particularly rich in intensely stained pigment granules (Fig. 7).

Because of the predominance of small nerve cells the presubiculum is denoted as parvocellular cortex or allocortical coniocortex (von Economo and Koskinas, 1925).

The presubiculum can be subdivided into three major territories, the presubiculum proper, the parasubiculum, and the transsubiculum. These reflect important influences maintained from the surrounding areas, which are laterally the allocortical subiculum, anteromedially the allocortical entorhinal region, and posteromedially the proisocortex.

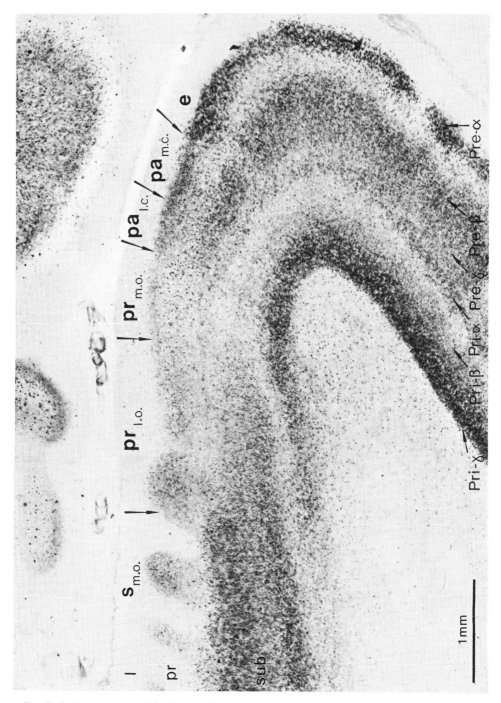

Fig. 7. Subiculum, presubiculum, and entorhinal region of man. Coronal section through upper parts of the parahippocampal gyrus at the latitude of the uncus. Note the interdigitation of subicular, presubicular, and entorhinal laminae which provide the basis for architectural parcellation of this territory. The layers are indicated along the *left* and the *lower margin* (*I* molecular layer, *pr* presubicular parvopyramidal layer, *sub* outer and inner pyramidal layer of the subiculum, *Pre-α* to *Pri-γ* laminae of the entorhinal region). Pigment preparation (800 μm)

5.2.1 The Proper Presubicular Subregion

Subjacent to the broad molecular layer is the parvopyramidal layer. It is already an almost solid plate. There follows a cell-sparse zone of variable breadth and in the lateral parts of the subregion irregularly shaped islands of large pyramidal cells. They are best characterized as belonging to the endogeneous layers of the subiculum.

Area presubicularis lateralis oralis (pr_{lo}) (Figs. 7, 12) displays an additional layer which is in continuation with the deep entorhinal layer Pri-γ. It becomes increasingly distinct as one approaches the medial border of the field.

Besides some indistinct remnants of Pri-β the area which follows caudally, *area presubicularis lateralis caudalis*, (pr_{lc}, Fig. 12) is devoid of entorhinal laminae.

The *area presubicularis posteropolaris* (pr_{pp}, Fig. 12), is also composed of only the parvopyramidal layer interspersed with clusters of CA1-pyramids and islands of subicular pyramidal cells.

The medial parts of the subregion are devoid of subicular pyramidal layers and show only the parvopyramidal layer and the entorhinal derivatives Pri-γ and Pri-β in *area presubicularis medialis caudalis* (pr_{mc}, Fig. 12) or additional to these layers rather thinly sketched Pre-γ and Pre-β in *area presubicularis medialis oralis* (pr_{mo}, Figs. 7, 12).

5.2.2 The Parasubicular Subregion

The parasubiculum is distinguished by a marked and abrupt change of the parvopyramidal layer: the dominant pyramidal cells show an increase in average cell size and keep larger distances between each other.

Area parasubicularis lateralis oralis (pa_{lo}) consists of the slightly changed parvopyramidal layer, islands of subicular pyramids and the entorhinal Pri-γ.

When passing over to *area parasubicularis medialis oralis* (pa_{mo}) the subicular pyramidal cells recede in favour of the entorhinal laminae Pre-β and Pre-γ.

Area parasubicularis lateralis caudalis (pa_{lc}, Figs. 7, 12) displays additionally a cell-rich Pri-β and a tenuous Pri-γ.

In *area parasubicularis medialis caudalis* (pa_{mc}, Figs. 7, 12) a thin layer of well-pigmented nerve cells appears which abuts almost directly on Pre-γ and is continuous with Pri-α.

5.2.3 The Transsubicular Subregion

The extension of the parasubiculum in the anteroposterior direction parallels that of the entorhinal region. As one proceeds occipital-wards, the entorhinal areas are substituted by isocortical fields. The multiform layer (VI) of the isocortex penetrates to a limited extent into the presubicular territory, establishing a transition zone designated the transsubiculum. It is an admixture of isocortical and allocortical laminae and belongs therefore to the periallocortex (H. Braak, 1978d).

The *area transsubicularis lateralis* (tr_l, Fig. 12) is mainly composed of the presubicular parvopyramidal layer and the multiform layer. In *area transsubicularis medialis* (tr_m, Fig. 12) an isocortical layer of small pyramids (II–IIIab) crops up, separating the molecular layer from the parvopyramidal layer. Cross-sections display the allocortical parvopyramidal layer slowly sweeping downwards (Fig. 10). The *area transsubicularis caudalis* (tr_c, Fig. 12) is wedged in between the retrosplenial region and the properistriate area. This field is almost totally formed of isocortical laminae except for a rather attenuated parvopyramidal layer which is localized immediately above layer IVb.

5.3 The Entorhinal Region

The entorhinal region is subjected to a remarkable increase in both extension and laminar elaboration as the phylogenetic scale is ascended (S. Rose, 1927; Sgonina, 1937; Stephan, 1956, 1960, 1961, 1966, 1975; Stephan and Andy, 1970). Influences of domestication lead to a more or less marked reduction in the extension of the region (Stephan, 1954; Kruska and Stephan, 1973).

In the human brain the entorhinal region spreads over both the gyrus ambiens and a considerable part of the parahippocampal gyrus (H. Braak, 1972a). Small wart-like elevations with shallow grooves in between can be found on parts of the surface of these gyri. These verrucae hippocampi (Klingler, 1948) mark the entorhinal region macroscopically and allow for its approximate delineation with the unaided eye.

The entorhinal cortex shows a particularly rich laminar differentiation. Each of the various layers displays gradually occurring changes in breadth. Some of the layers disppear within the limits of the region. The alterations of the lamination pattern provide a basis for subdivision of the region. A narrow cell-sparse plexiform layer (lamina dissecans) permits distinction between the external (Pre) and internal main stratum (Pri).

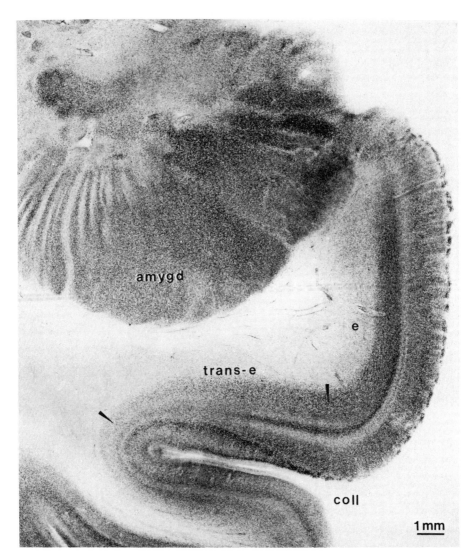

Fig. 8. Low-power view of the entorhinal region of man. Coronal section through anterior parts of the entorhinal region at the latitude of the corpus amygdaloideum (*amygd*). Here the proper entorhinal subregion (*e*) covers almost totally the free surface to the parahippocampal gyrus. The transentorhinal subregion (*trans-e*) is buried in the depth of the collateral sulcus (*coll*). The intensely stained islands of Pre-α coalesce in the vicinity of the collateral sulcus and run obliquely through the outer main stratum. Hence within the transentorhinal zone of transition allocortical and isocortical layers are intimately indented. The borders of the transentorhinal subregion are marked by *triangles*. Pigment preparation (1000 μm)

5.3.1 The Proper Entorhinal Subregion

The *lateral central entorhinal field* (e_{ce1}, Figs. 9, 12) reveals the culmination point of laminar elaboration. The complete set of entorhinal laminae can be recognized:

The *molecular layer* is broad. It is composed of an external glial layer, a zone rich in myelinated fibres (sublamina supratangentialis) and a cell- and fibre-sparse lower part (sublamina tangentialis).

The outer main stratum consists of three laminae: Pre-α, Pre-β, and Pre-γ (M. Rose, 1927a,b).

Layer Pre-α is distinguished by islands or lines of nerve cells embedded in the neuropil of the molecular layer. The islands, formed of medium-sized to large nerve cells, vary in size and shape (Figs. 7, 8, 9, 10, 11). Cell stains reveal polygonal perikarya with coarse Nissl granules and relatively large nuclei. The cell bodies contain a fair number of intensely stained pigment granules which are concentrated in one part of the cytoplasm. Numerous dendrites, each of approximately the same length, are generated from the cell body by way of cone-shaped proximal stems. The dendrites give off side branches in great numbers and are richly invested with spines. They can issue from any point of the cell body and form in this way a globular dendritic domain. The stout axon descends and reaches the white matter after having issued some 6 to 10 collaterals (Ramón y Cajal, 1909; Lorente de Nó, 1933; H. Braak et al., 1976) (Fig. 2 A).

Besides this multipolar type of nerve cells, only a few large stellate cells occur of either the non-pigmented or the pigment-laden variety. They generate a few thin, far-reaching, and infrequently bifurcating dendrites with smooth and in places varicose contour. The local axon arborizes profusely (Lorente de Nó, 1933; H. Braak et al., 1976).

Fine radially oriented bundles of myelinated fibres extend almost up to layer Pre-α (Fig. 11). It appears nonetheless doubtful to characterize the entorhinal cortex as being supraradiate (Sgonina, 1937) since allocortical laminae cannot be homologized with isocortical ones. Usage of the characteristics of the radiate bundles should therefore be confined to the isocortex. If with anything of all, the superficial lamina Pre-α has traits in common with the isocortical layer IIIc. Considering this, the entorhinal cortex would actually be of the euradiate type.

Immediately subjacent to Pre-α there is a cell-sparse band of tangentially adjusted myelinated fibres (Fig. 11).

Layer Pre-β, which is mainly formed of slender pyramidal cells, follows. The triangular or even spindle-shaped cell bodies give rise to a thin apical dendrite without ramifications piercing Pre-α. The short root-shaped basal process bursts into a rich bushel of dendrites heading more

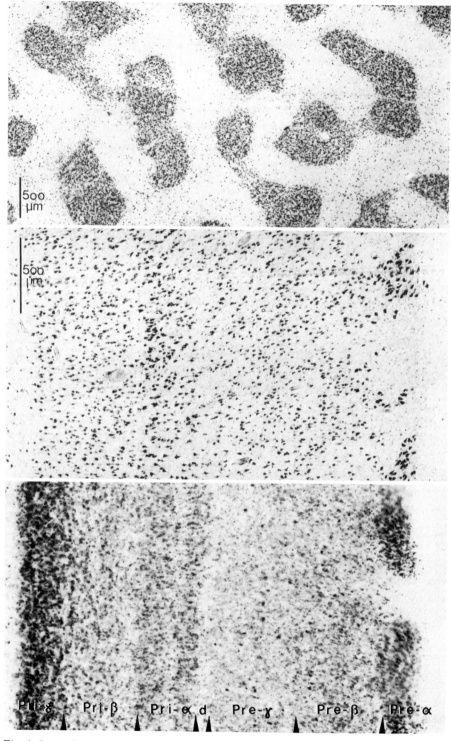

Fig. 9. Legend see p. 41

or less straight downwards (tassel cells: von Economo and Koskinas, 1925). These cells store only a moderate amount of finely grained pigment (Figs. 9, 11).

Layer Pre-γ appears as a cell-sparse zone where slender spindle-shaped and rather poorly pigmented pyramids prevail (Figs. 9, 11).

Immediately underneath the clear-cut *lamina dissecans,* which is devoid of nerve cells and rich in tangentially adjusted myelinated fibres (Fig. 11), there follows *layer Pri-α* which is mainly composed of well-formed pyramids. In the central fields of the entorhinal region, the layer is split into a cell-rich Pri-αα, a cell-sparse Pri-αβ, and again a cell-rich Pri-αγ (Fig. 9). The cell bodies of the pyramids are richly endowed with pigment.

The dominating constituents of the *layer Pri-β* are slender pyramidal cells of small size giving rise to a thin apical dendrite and some delicate basal ones. Initially, the axon descends but soon bends back, piercing the overlying strata. It splits up into terminal ramifications within the superficial laminae. These cells are almost barren of lipofuscin granules. They plunge into a thick plexus of myelinated horizontal fibres (Figs. 9, 11).

The nerve cells inhabiting *layer Pri-γ* display differently shaped cell bodies and peculiarities of their dendritic arbours. The majority of these cells are richly endowed with lipofuscin deposits so that Pri-γ stands out in pigment preparations (Figs. 9, 11). In Nissl preparations, layers Pri-β and Pri-γ are often difficult to distinguish from each other (Fig. 11).

In places the entorhinal cortex shows a considerable number of spindle-shaped nerve cells irregularly disseminated throughout wide parts to the white substance. This accumulation of white matter neurons is denoted as *lamina cellularis profunda* (H. Braak, 1972a).

The basic pattern just described is subjected to incessant local variations. Based on cytoarchitectonics, Rose (1927a,b, 1935) distinguishes 23 fields. Sgonina (1938) adopts the mapping of Rose. Unfortunately, it is often unclear from the descriptions of both authors which criteria they use for delineation.

Pigment preparations permit clear distinction between the various cellular layers and therefore provide a basis for a reliable parcellation of the entorhinal cortex (H. Braak, 1972a).

The disappearance of lamina Pre-γ indicates the borderline between *area entorhinalis centralis lateralis* and *medialis* (e_{cel}, e_{cem}, Fig. 12).

◀ **Fig. 9.** Entorhinal region of man. *Upper third* Islands of layer Pre-α. Pigment preparation cut tangentially to the cortical surface (800 μm). *Middle* and *lower third* Nissl and pigment preparation (15 and 800 μm) of area entorhinalis centralis lateralis. Note the splitting of layer Pri-α which is a hallmark of the central entorhinal fields. The layers are indicated along the *lower margin*

In *area entorhinalis interpolaris medialis* (e_{im}, Fig. 12), the tripartition of Pri-α is absent, Pri-α appears rather as a solid line. Lamina Pre-γ is lacking.

The deep layers Pri-α and Pri-γ unite to form a broad band in *area entorhinalis interpolaris oralis* (e_o, Fig. 12).

Area entorhinalis interpolaris lateralis (e_{il}, Fig. 12) differs from the central fields in that it shows a uniform Pri-α. In contrast to the medial interpolar field it has a tripartite outer main stratum.

The gyrus ambiens is covered by three areas which altogether show an outer main stratum composed of only Pre-α and Pre-β. The *central field* (ga_c, Fig. 12) displays clear separation of Pri-α and Pri-γ by a clear stripe (Pri-β). Additionally, the field is endowed with a particularly cell-rich lamina cellularis profunda.

In the *lateral field* (ga_l, Fig. 12) the deep laminae Pri-α, Pri-γ, and the lamina cellularis profunda appear considerably attenuated.

The *oral field* (ga_o, Fig. 12) shows confluence of Pri-α and Pri-γ.

5.3.2 The Transentorhinal Subregion

The lateral circumference of the entorhinal region abuts on the temporal proisocortex. There is no clear-cut border between the allocortex and isocortex but by contrast an expanded territory where allocortical and isocortical laminae meet in a zone of mutual interdigitation. This stretch of cortex accordingly is part of the periallocortex. It is particularly expanded in the human brain. Subhuman primates show a markedly less extended transition zone.

The leading features of the transentorhinal subregion are shown in Fig. 12. The parcellation is based only on the pigment preparations since

---

Fig. 10. *Upper third* Transsubicular subregion of man. At *left hand margin* the molecular layer is followed by the parvopyramidal presubicular layer (*prs*) and the multiform layer (area transsubicularis lateralis). As one proceeds *towards the right hand margin* irregularly spaced clouds of the third isocortical layer (*III*) become more and more prominent. The parvopyramidal layer sweeps downwards and lies finally between the isocortical layers III and VI (area transsubicularis medialis). Nissl preparation (15 μm). Coronal section cut at the latitude of the splenium corporis callosi. *Middle third* Transentorhinal subregion of man. The pigment preparation shows clearly the confluence of Pre-α islands and the wedge-shaped isocortical laminae II and III which separate Pre-α from the molecular layer. Coronal section (800 μm). *Lower third* The same subregion at the same latitude as seen in the Nissl preparation (15 μm). Delineation of the various laminae is extremely difficult. The course of the Pre-α derivatives is indicated by a *thin interrupted line*

The Transentorhinal Subregion

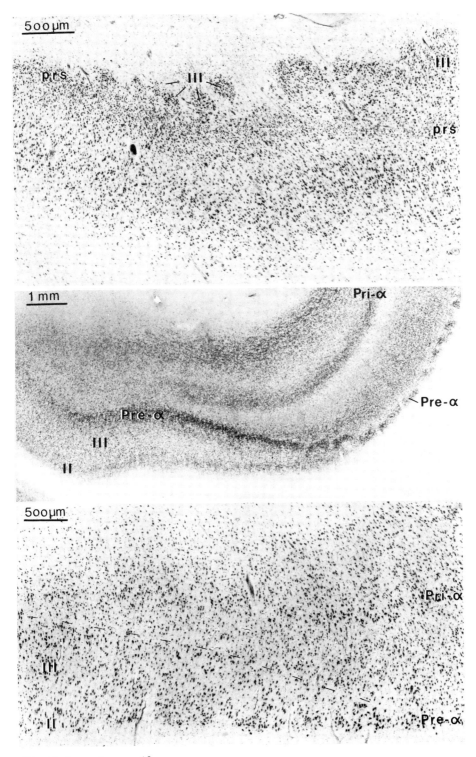

Fig. 10. Legend see p. 42

the peculiar construction of the transition zone is not displayed to advantage in routine Nissl or myelin preparations, a fact that has often led to the assumption that the allocortex is gradually and continually transformed into the isocortex (von Economo and Koskinas, 1925). Figures 8 and 10 show the interdigitating laminae which clearly do not amalgamate.

Pre-α is generally an already consolidated layer. In *area transentorhinalis medialis* (e_{trm}, Figs. 8, 10, 12) it begins to sweep downward following an oblique course through the outer laminae. The *intermediate transentorhinal field* begins when Pre-α passes the border between Pre-β and Pre-γ. Isocortical layers (II + III) fill up the wedge-shaped space between Pre-α and the molecular layer. The strongly pigmented Pre-α extends far into the neighbouring cortex as a rather tenuous plate which is finally located between the third and the fourth isocortical layer (*area transentorhinalis lateralis*).

Pigment-laden Pre-α derivatives and pigment laden IIIc-pyramids have some traits in common. The entorhinal fields as allocortical "magnopyramidal" region.

On the way downwards, the multipolar nerve cells of Pre-α become regularly aligned. The cells gradually adopt the features of typical pyramids, a fact which corroborates the classification of the superficially located Pre-α neurons as modified pyramids (H. Braak et al., 1976).

It seems worthwhile to point to the fact that the richly pigmented Pre-α pyramids gain more and more the traits of large pigment-laden IIIc-pyramids. Nerve cells of this peculiar type can rarely be encountered within the telencephalic cortex but occur in only a few highly refined areas such as the temporal or frontal speech areas (see Chap. 7.2.4 and 7.4.3). The structural similarities suggest a close relationship between the main constituents of Pre-α and the pigment-laden IIIc-pyramids of the isocortex.

In this light the entorhinal region can be considered a refined "association" centre of the allocortex. Despite its connections with the prepiriform and periamygdalear cortex (Cragg, 1961; Powell et al., 1965; Valverde, 1965; Shute and Lewis, 1967; Price and Powell, 1971; van Hoesen et al., 1972; van Hoesen and Pandya, 1973; Price, 1973; Krettek and Price, 1977b), it seems rather unlikely that the entorhinal region serves only as

---

Fig. 11. Area entorhinalis interpolaris lateralis. Coronal sections successively cut from ▶ the same block of tissue. *Upper third* Nissl preparation (30 μm). *Middle third* Myelin preparation (100 μm). *Lower third* Pigment preparation (800 μm). Note the improved demonstration of the various entorhinal laminae in the pigment preparation as compared to the Nissl preparation. In particular the layers Pre-γ and Pri-β cannot easily be delineated in the Nissl preparation

The Transentorhinal Subregion

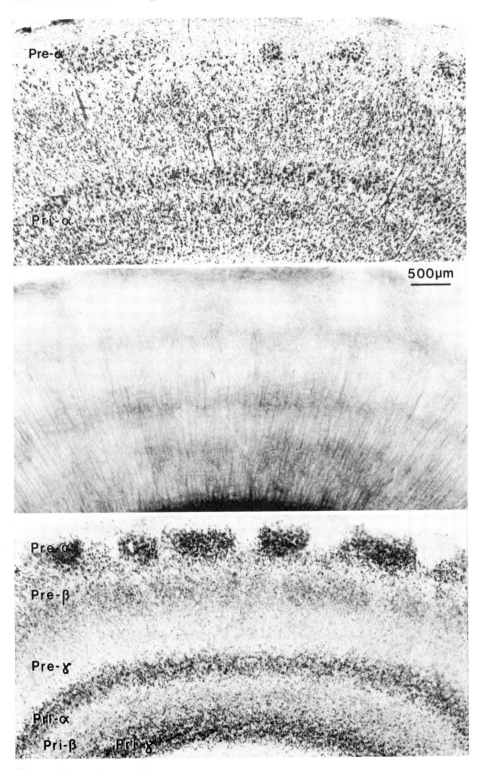

Fig. 11. Legend see p. 44

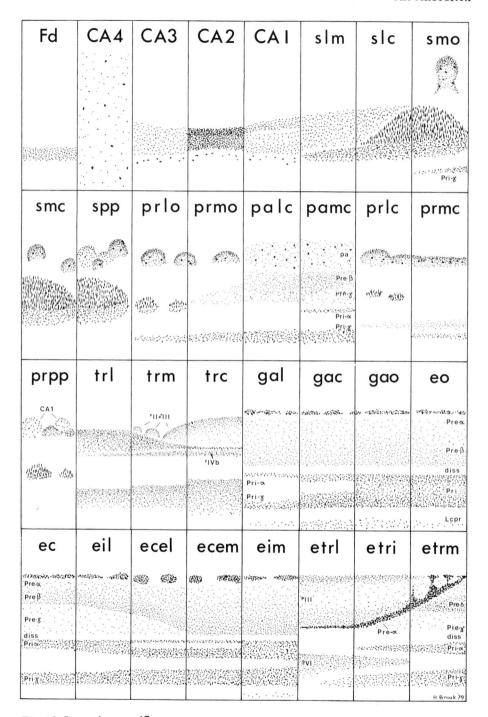

Fig. 12. Legend see p. 47

The Transentorhinal Subregion

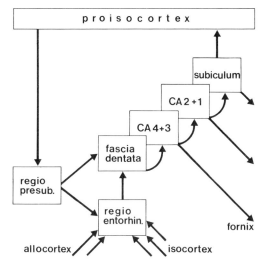

Fig. 13. Simplified scheme of the main connections established between the various parts of the allocortex described

◄ Fig. 12. Diagram showing the characteristic lamination pattern of the various allocortical areas in pigment preparations.

Fd	Fascia dentata
CA1–CA4	Sectors of the cornu ammonis
slm	Area subicularis lateralis marginalis
slc	Area subicularis lateralis centralis
smo	Area subicularis medialis oralis
smc	Area subicularis medialis caudalis
spp	Area subicularis posteropolaris
prlo	Area presubicularis lateralis oralis
prmo	Area presubicularis medialis oralis
palc	Area parasubicularis lateralis caudalis
pamc	Area parasubicularis medialis caudalis
prlc	Area presubicularis lateralis caudalis
prmc	Area presubicularis medialis caudalis
prpp	Area presubicularis posteropolaris
trl	Area transsubicularis lateralis
trm	Area transsubicularis medialis
trc	Area transsubicularis caudalis
gal	Area entorhinalis gyri ambientis lateralis
gac	Area entorhinalis gyri ambientis centralis
gao	Area entorhinalis gyri ambientis oralis
eo	Area entorhinalis oralis
ec	Area entorhinalis caudalis
eil	Area entorhinalis interpolaris lateralis
ecel	Area entorhinalis centralis lateralis
ecem	Area entorhinalis centralis medialis
eim	Area entorhinalis interpolaris medialis
etrl	Area transentorhinalis lateralis
etri	Area transentorhinalis intermedia
etrm	Area transentorhinalis medialis

a centre of olfaction. On the contrary its main input comes from the surrounding isocortical fields (Jones and Powell, 1970e; Price and Powell, 1971; van Hoesen et al., 1972; van Hoesen and Pandya, 1973; Seltzer and Pandya, 1974). Via the perforant path, it projects to the hippocampal formation (Hassler, 1964; Andersen et al., 1971; Hjorth-Simonsen, 1972; Segal and Landis, 1974). The entorhinal region appears therefore rather as an important link between the isocortex and the hippocampal formation (Fig. 13).

6 The Proisocortex

The temporal lobe contains not only the most important part of the allocortex but is also richly endowed with extended periallocortical areas which incompletely surround the core, thus forming a transition to the neighbouring isocortex (transsubicular and transentorhinal areas with a number of adjoining temporal fields; H. Braak, 1978c).

As one follows the allocortical derivatives on to the ridge of the corpus callosum both the allocortical and periallocortical structures diminish markedly in extension. The adjacent proisocortex by contrast comes into prominence and spreads over considerable parts of the cingulate gyrus.

Two regions can be distinguished which differ markedly from each other: the retrosplenial region posteriorly and the anterogenual region anteriorly. Both regions are heterotypic in character. The retrosplenial region is for the most part formed of parvocellular cortex as opposed to the anterogenual region, which is magnocellular. Both regions are pulled far apart from each other and reach their greatest expansion either in front of the genu or back of the splenium corporis callosi (H. Braak, 1979c,d) (Figs. 13, 15, 34).

6.1 The Retrosplenial Region

Most authors agree in the subdivision of the retrosplenial region into a "hypergranular" part which is in continuation with the supracallosal allocortex and an "agranular" one mediating to the isocortex. The granularized zone can be further subdivided into periallocortical and proisocortical parts. Table 1 gives a synopsis of the nomenclature used by different authors for designation of the retrosplenial areas.

The retrosplenial core fields rarely extend on to the free surface of the cingulate gyrus. Their outlining in the maps of Brodmann (1909) and von Economo and Koskinas (1925) does not coincide with their real extension.

The *induseum griseum* covering the corpus callosum is medially formed of patchy remnants of the fascia dentata which abut on strands of ammonshorn constituents. Lateral-wards follows a more or less eroded deep layer

Table 1. Synopsis of nomenclature concerning the retrosplenial region of man according to various authors

Brodmann, 1908, 1909 (cyto) Stephan, 1975 (cyto, myelo)	ectospl. (26)	retrospl. gran. (29)		retrospl. agran. (30)	
von Economo and Koskinas, 1925; von Economo, 1926, 1927 (cyto)	retrospl. gran. inf. (LE2)	retrospl. gran. sup. (LE1)		retrospl. agran. (LD)	
Rose, 1928, 1935 (cyto, myelo)	retrospl. gran.med. (RSgα)	retrospl. gran.interm. (RSgβ)	retrospl. gran.lat. (RSgγ)	retrospl. agran. (RSag)	
Braak, H., 1979c,d (cyto, myelo, pigmento)	ectospl. (periallo)	retrospl. lat. (proiso)	retrospl. intermed. (proiso)	retrospl. med. (proiso)	paraspl. (iso)

of internal subicular pyramids (Fig. 14). These allocortical structures represent the supracommissural parts of the hippocampal formation (Stephan, 1975).

The adjoining *periallocortical ectosplenial field* (Figs. 14, 15) hardly shows a clear lamination. The molecular layer is filled with myelinated fibres. An outer main stratum (III) can just be distinguished from an inner one which seemingly consists of only the multiform layer (VI).

The following proisocortical fields, *area retrosplenialis lateralis et intermedia,* are frequently regarded as the "heart" of the heterotypical region (Stephan, 1975). The border with the foregoing field is sharply drawn.

The molecular layer is broad and still contains a particularly thick plexus of tangentially adjusted myelinated fibres.

The breadth of the subjacent band of cells exceeds by far that of the inner layer. It is composed of the larger constituents of the pyramidal layer (III) with small pyramids interspersed. A granular layer (IV) is absent. The different varieties of both the pyramidal and the stellate cells are relatively small, a fact which mainly contributes to the granularized character of the cortex (Vogt, 1976). Myelin preparations reveal an extremely

Fig. 14. *Upper half* Location of the retrosplenial region and diagrammatic representation of its various areas in the Nissl, myelin, and pigment preparations. *Lower half* Frontal sections through posterior parts of the cingulate gyrus. The boundaries between areas are marked by *triangles*. Pigment preparation (800 μm) above, and myelin preparation (100 μm). *a* Allocortex; *Be* and *Bi* outer and inner line of Baillarger; *Cc* corpus callosum; *es* area ectosplenialis; *ps* area parasplenialis; *rsi* area retrosplenialis intermedia; *rsl* area retrosplenialis lateralis; *rsm* area retrosplenialis medialis; *Te* and *Ti* outer and inner tenia

The Retrosplenial Region

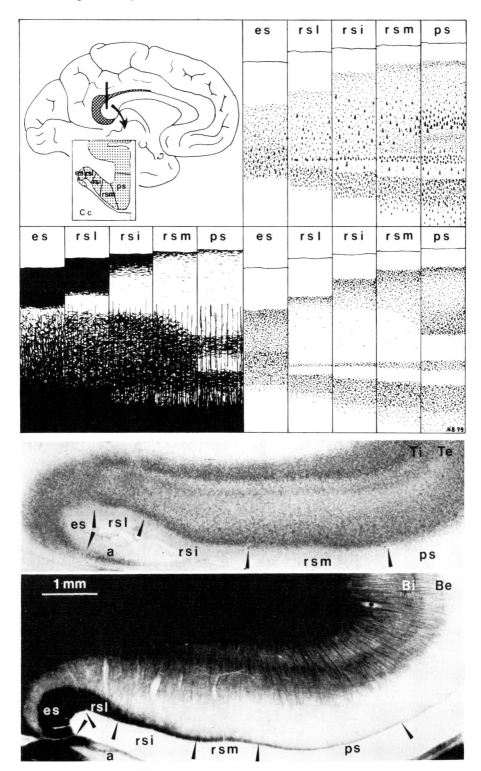

Fig. 14. Legend see p. 50

broadened outer line of Baillarger. It shows blurred upper and lower borders, probably due to the fact that there is no clear-cut granular layer. The broad plexus lies within the limits of the pyramidal layer. In pigment preparations, this outer cellular band is particularly distinguished by its pallor. Narrow superficial parts with pigmented pyramids are followed by an extremely broad external tenia which covers major parts of the outer main stratum. The amount of the pigment stored within the nerve cells decreases gradually as the cortex is descended, a rather uncommon feature which occurs in only a few cortical fields.

Layer Va is almost lacking and is represented only by a tenuous though sharply delimited band of pigmented nerve cells. The layer becomes increasingly rarefied as one approaches the periallocortex. Accordingly, the retrosplenial cortex is externocrassior, in this comparable only to sensory core fields such as the striate area or the postcentral coniocortex. Rose (1928) on the contrary designates the lower reaches of the outer main stratum as Va. The cortex would then be extremely internocrassior. Pigment preparations by contrast clearly reveal the existence of an only thinly sketched ^{P}Va which is furthermore in continuation with a better developed ^{P}Va of the adjoining isocortex. It seems moreover rather unlikely that a pronouncedly parvocellular cortex should be endowed with a particularly broad ganglionic layer.

Layer Vb is a narrow cell-sparse stripe. In pigment preparations it appears as a pale line, the internal tenia. The marked breadth of the external tenia as opposed to the thin internal one is one of the hallmarks of the retrosplenial cortex.

The multiform layer (VI) is continuous with the deep layer found already in the ectosplenial field. It shows clear-cut upper and lower boundaries. The border towards the periallocortex is marked by the sudden appearance of richly pigmented constituents. In pigment preparations, ^{P}VI is the most intensely stained layer of the retrosplenial cortex. Both retrosplenial areas show an almost astriate character because of the high myelin content of the deep laminae.

The intermediate retrosplenial area differs from the lateral one in that the molecular layer is less rich in myelinated fibres. The externoteniate characteristic is a trifle less expressed. Also ^{P}Va — though still a narrow band — is a better recognizable layer in the intermediate field.

Fig. 15. Retrosplenial region of man. Frontal sections successively cut from the same ▶ block of tissue for comparison of the architectonic features revealed by the pigment, myelin, and cell staining methods. Note the gradual attenuation of the outer tenia within the limits of the medial retrosplenial area. Pigment preparation (800 µm), myelin preparation (100 µm), Nissl preparation (15 µm). *Be* and *Bi* outer and inner line of Baillarger; *ps* area parasplenialis; *rsi* area retrosplenialis intermedia; *rsm* area retrosplenialis medialis; *Te* and *Ti* outer and inner tenia. *IIIc, V, VI* isocortical layers

The Retrosplenial Region

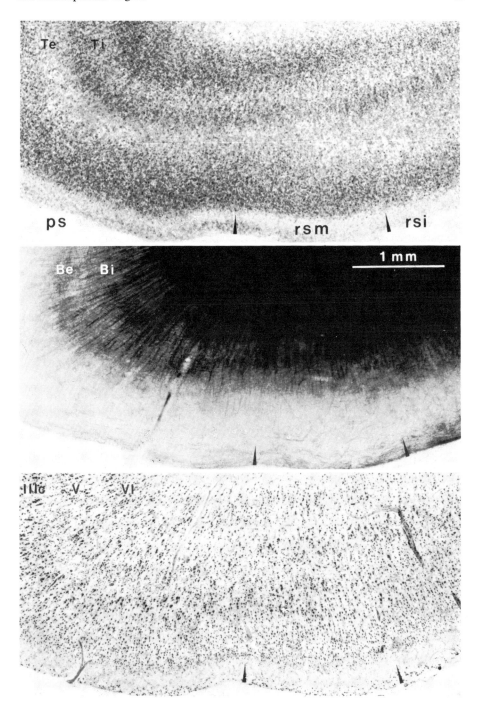

Fig. 15. Legend see p. 52

Area retrosplenialis medialis is still of the externoteniate type. There is a considerable numerical decrease of myelinated fibres in the molecular layer. The average size of the outer main stratum constituents is increased. A sublayer IIIab mainly formed of small pyramids becomes distinguishable from a sublayer IIIc. Cytoarchitecturally, the area is already classified with the "agranular" part. There is nevertheless an incipient granular layer (IV) recognizable. As one proceeds medially, the pigmented outer zone (^{P}III) gradually increases in breadth at the expense of the external tenia. This corresponds with a similar attenuation of the outer line of Baillarger.

^{P}Va is a step broader than in the intermediate field. ^{P}VI is split into a darker upper (^{P}VIa) and a lighter lower zone (^{P}VIb). The substriate zone (6) appears, which gives the cortex a unitostriate character.

As the cortex is traced forwards out of the sulcus into the free surface of the cingulate gyrus, the extremely externoteniate character vanishes. The only weakly externoteniate *area parasplenialis* (Figs. 14, 15) mediates to the equoteniate parietal cortex. The parasplenial field belongs already to the mature isocortex. It can be considered a belt area to the retrosplenial core fields.

The main difference to the foregoing fields is that the pyramidal layer (^{P}III) shows the common pattern of pigmentation, i.e., its main constituents store the greater amounts of lipofuscin granules the deeper their position is. The lower border of ^{P}III is clear-cut. A granular layer (IV) appears. The outer line of Baillarger is narrow with a relatively sharp upper border.

^{P}Va is a well-developed layer. An intrastriate band (5a) becomes recognizable. Both the internal tenia and the multiform layer remain unchanged.

Summing up, the parvocellular retrosplenial core fields show a marked preponderance of the outer main stratum and a clear externocrassior and externoteniate characteristic. They are averagely poor in pigment (typus clarus) and rich in myelin fibres (typus dives). The parasplenial belt area forms a transition to the equoteniate parietal isocortex. It is only weakly externocrassior and externoteniate. The homotypical field shows an already average pigment and myelin content.

Remarks Concerning the External Tenia. Extremely *externoteniate* areas are rarely encountered in the telencephalic cortex. Within the isocortex, they occur only within the limits of the visual, acoustic, somatosensory, and somatomotor cores.

Often the outer tenia corresponds well to a broad outer band of Baillarger. But there are also areas with an expanded outer tenia displaying an insignificant external Baillarger.

The lines of Baillarger are mainly formed of myelinated axon collaterals of pyramidal cells (Clark, Le Gros and Sunderland, 1939; Braitenberg, 1962, 1974, 1978; Fisken et al., 1975). Pyramids which lie within a thick plexus of axonal endramifications generally remain almost devoid of lipofuscin also as age advances, provided that the plexus is formed of pyramidal cell axon collaterals or of thalamo-cortical or amygdalo-cortical fibres. The fibre plexus may be partly unmyelinated and therefore only partly visualized in myelin preparations. The total breadth of the axonal plexus would then only be displayed in pigment preparations. Thalamocortical fibres ramify preferably in the lower reaches of the outer main stratum. Areas known to receive a massive thalamic input such as the isocortical core fields, accordingly, appear outstandingly externoteniate.

It has long been thought that the so-called "limbic lobe" which includes both anterior and posterior territories of the cingulate gyrus receives special input from the anterior nuclei of the thalamus (Yakovlev et al., 1960). Recent investigations by contrast give strong support to the assumption that it is only the retrosplenial cortex which is under the influence of the anterior thalamic nuclei and is additionally endowed with a wealth of fibres originating from the surrounding isocortical areas (Cuénod et al., 1965; MacLean and Creswell, 1970; Vogt et al., 1979) and the subiculum (Rosene and van Hoesen, 1977). These findings are in broad agreement with the fact that within the cingulate gyrus the extremely externoteniate character is confined to the retrosplenial core fields.

6.2 The Anterogenual Region

Traced anteriorly, the retrosplenial cortex extends approximately to the latitude of the mamillary bodies. From here on it is relatively abruptly replaced by a magnocellular and agranular cortex. This considerably expands as one proceeds frontal-wards. The broad contact between the retrosplenial and anterogenual regions which is generally found in the lower mammals is condensed to a narrow zone close to the corpus callosum in the higher primates and in man (Rose, 1928).

The anterogenual region stretches out on to the free surface of the cingulate gyrus covering a crescent-shaped territory in front of the genu corporis callosi (Fig. 16). Characteristic features permit easy delineation of the anterogenual areas in pigment preparations. In myelin preparations by contrast difficulties arise in distinguishing these areas from adjoining paralimbic and subgenual fields. Extended parts of the anterior cingulate gyrus appear also more or less uniformly agranular and magnocelluar in

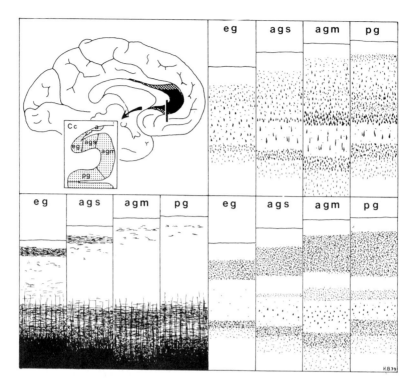

Fig. 16. Location of the anterogenual region (free surface of anterogenual areas: *black*, of the paragenual field: *dotted*) and diagrammatic representation of the areas in the Nissl, myelin, and pigment preparations. *a* Allocortex; *agm* area anterogenualis magnoganglionaris; *ags* area anterogenualis simplex; *Cc* corpus callosum; *eg* area ectogenualis; *pg* area paragenualis

the Nissl preparations, facts which possibly account for the only rough correspondance between the various cyto- and myeloarchitectonic maps (Figs. 35–43) and the present pigmentoarchitectonic parcellation (Figs. 14, 16, 34). The anterogenual region covers only parts of the subregio praecingularis (Brodmann, 1908, 1909) or the infraradiate region (Vogt, 1910; Vogt and Vogt, 1919; Rose, 1928; Strasburger, 1937; Stephan, 1975).

Fig. 17. Frontal section through subcallosal parts of the anterogenual region. Note the ▶ conspicuous population of pigment-laden nerve cells within the broad internal tenia. Pigment preparation (800 μm). *a* Allocortex; *agm* area anterogenualis magnoganglionaris; *ags* area anterogenualis simplex; *C.call.* corpus callosum; *eg* area ectogenualis; *pg* area paragenualis; *Te* and *Ti* outer and inner tenia; $pIII$, pVa, pVI layers of pigmented nerve cells

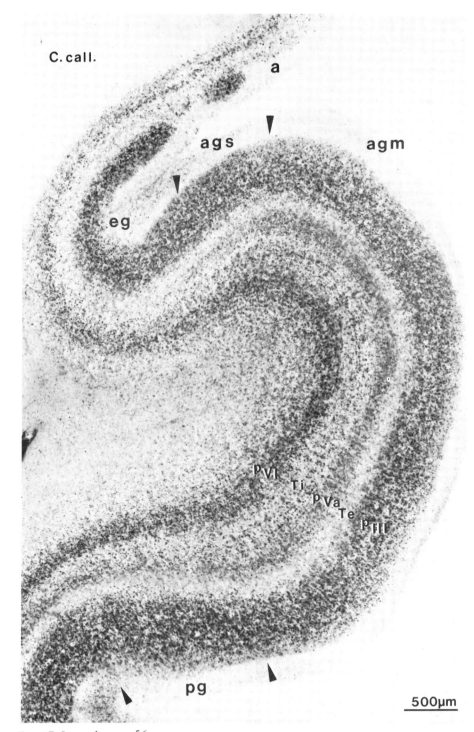

Fig. 17. Legend see p. 56

The anterogenual core fields correspond to only anterior parts of field 24 and posterior parts of field 33 (Brodmann, 1908, 1909), parts of the fields 17–24 (Vogt, 1910), LA_1 and LA_2 (von Economo and Koskinas, 1925), or IRb and IRc (Rose, 1928).

The allocortical *induseum griseum* outside the anterogenual region is still formed of nerve cells which can be recognized as derivatives of the cornu ammonis. They maintain close quarters with a flattened lamella of small nerve cells which as to its structural features is a remnant of the deep layer of subicular pyramids. This layer, in turn, is often continuous with the isocortical lamina ^{P}VIb.

An ateniate or propeateniate stretch of cortex is missing or only poorly developed. The *ectogenual field* which might still be considered a part of the periallocortex is unitoteniate. The molecular layer is rich in myelinated fibres. The outer cellular layer (^{P}III) is densely pigmented. It is followed by a broad stripe which results from the coalescence of both teniae, and a bipartite multiform layer. The outer cellular lamina is devoid of a corpuscular layer. The united teniae harbour numerous sparsely pigmented nerve cells which are evenly disseminated and can be considered forerunners of layer ^{P}Va. Already close to the allocortical remnants, the multiform layer is split into a densely pigmented upper part (^{P}VIa) and a modestly pigmented lower one (^{P}VIb). ^{P}VIa is not in as a tight connection with the subicular band as is ^{P}VIb.

The following biteniate fields, *area anterogenualis simplex* and *magnoganglionaris,* can be considered the "heart" of the magnocelluar anterogenual region. The fields belong to the proisocortex. In Golgi preparations, their constituents display numerous primitive features (Schierhorn et al., 1972a; Schulz and Schönheit, 1974; Michalski et al., 1976; Schönheit and Schulz, 1976; Schulz et al., 1976). The cortex is nevertheless totally composed of isocortical laminae. The elaboration of the anterogenual core culminates in the magnoganglionic field which is therefore described in detail.

The molecular layer is broad and contains only a sparse number of myelinated fibres.

The pyramidal layer (III) is richly stocked with well-pigmented pyramids. Both the upper and the lower border of the pyramidal layer (^{P}III) are sharply drawn in pigment preparations. The external tenia is a pallid stripe filled with nerve cells of roughly the same size and shape as those

---

Fig. 18. *Left* Lamination of the magnoganglionic anterogenual field (Nissl preparation, ▶ 15 µm). *Right* Frontal sections successively cut from the same block of tissue for comparison of the architectonic features of the magnoganglionic anterogenual field as revealed by pigment, myelin, and Nissl preparations. Pigment preparation (800 µm), myelin preparation (100 µm), Nissl preparation (15 µm). *Te* and *Ti* outer and inner tenia; ^{P}*III*, ^{P}*V*, ^{P}*VI* layers of pigmented nerve cells; *III, Va, Vb, VI* isocortical layers

The Anterogenual Region 59

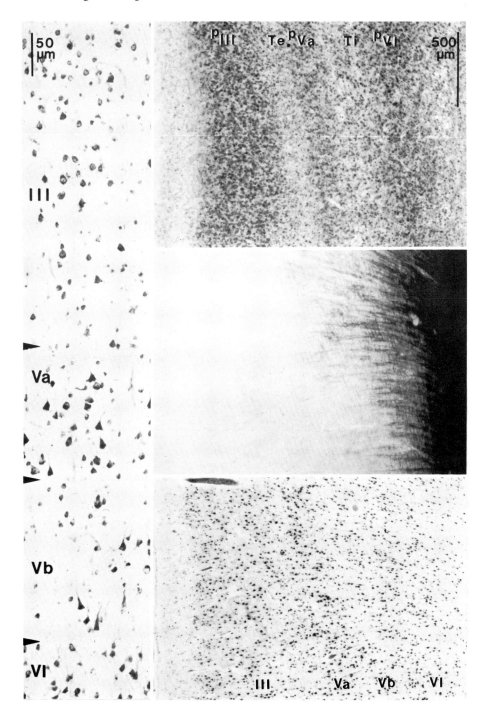

Fig. 18. Legend see p. 58

found in ^{P}III with some smaller ones interspersed. The perikarya of these cells are devoid of pigment deposits. Myelin preparations show a pallid appearance of the outer laminae. The external line of Baillarger cannot be outlined. The granular layer (IV) is lacking.

The following layer Va is a narrow but clearly recognizable band consisting of densely placed large pyramids which contain only a small number of feebly tinged pigment granules.

Layer Vb shows a widening of the average interval between the individual pyramids. It is remarkably broad and contains a plexus of myelinated horizontal fibres. Thin bundles of radially oriented myelinated fibres extend up to the fifth layer. The infraradiate character thus achieved is one of the hallmarks of the anterogenual fields. The inner tenia (^{P}Vb) shows clear-cut boundaries and is in places more than twice as broad as the external tenia. It appears darkened to some extent due to its content of numerous large and particularly slender pyramidal cells which amass lipofuscin granules in an aggregate. This unusual type of pigmentation marks the large nerve cells, which on account of their location within the pallid internal tenia catch the eye all the more.

These large pyramidal cells are special constituents of the anterogenual core fields. In Nissl preparations they appear as spindles with extended perikarya (stick cells or corkscrew cells: von Economo and Koskinas, 1925; Ngowyang, 1932; Sanides, 1962; Stephan, 1964, 1975). Until now, cells of this type have only been described in the primate brain. They are particularly numerous and distinct in the brain of man (Rose, 1928).

The stout apical dendrite of these cells reaches the molecular layer and bursts into its terminal arborization. The soma issues only a small number of short basal dendrites. The axon contributes to the cingulum (Stephan, 1975).

The distinctive pattern of pigmentation and the location in sublayer Vb allows one to classify this type of pyramidal cells with the class of Betz cells (see Chap. 2.2 and 7.4.1).

In Nissl stains there is a clear preponderance of the ganglionic layer in comparison to IIIc (internocrassior character).

The multiform layer (VI) is split into a cell-dense VIa and a cell-sparse VIb. The upper reaches of the multiform layer (^{P}VIa) appear as well-pigmented clear-cut band. The layer contains a great number of myelinated fibres.

The *area anterogenualis simplex* shows up less pronouncedly the features of the magnoganglionic field. The cortex as a whole is smaller. A rich plexus of myelinated fibres within the molecular layer points to the vicinity of the ectogenual field. Both teniae are less well delimited than in the magnoganglionic field. ^{P}Va is only thinly sketched. ^{P}Vb is less richly endowed with spindle-shaped pigment-laden pyramids. The multiform layer remains unchanged.

The *area paragenualis* which accompanies the magnoganglionic field is only weakly internoteniate, thereby forming a transition to the equoteniate areas of the frontal lobe.

The molecular layer is almost devoid of myelinated fibres. There follows an incipient corpuscular layer (II) and a pyramidal layer (III) which is broader than in the magnoganglionic field. There appears also a tenuous granular layer (IV). The myelin content of the outer laminae remains very low.

The breadth of both the ganglionic and the multiform layer is reduced. Sublayer Vb in particular is attenuated, a fact which is also shown by a corresponding decrease in the width of the internal tenia. The layer is only modestly endowed with pigment-laden slender pyramids. The multiform layer (VI) is unchanged in its composition.

In sum, the leading features of the anterogenual core are first the occurrence of large pigment-laden pyramids in ^{P}Vb, and second the extremely internoteniate characteristic. The fields are furthermore internocrassior. The heterotypical agranular and magnocellular cortex is rich in overall pigmentation (typus obscurus) and poorly endowed with myelinated fibres (typus pauper). The paragenual belt area forms a transition to the equoteniate frontal isocortex. It is only weakly internocrassior and internoteniate. It is still magnocellular, rich in pigment and poor in its myelin content.

Remarks Concerning the Internal Tenia. Marked prevalence of the internal tenia is rarely encountered in the telencephalic cortex. In all probability, the outstanding breadth of the inner tenia results from the presence of a thick plexus of axonal endramifications, as has been considered already. The plexus is presumably composed not only of axon collaterals of the densely placed Va-pyramids but also of extrinsic afferent fibres. The anterior cingulate cortex receives only a few afferents from other cortical areas (Vogt et al., 1979). Recent investigations have shown that the amygdalocortical projection preferably spreads out in the ganglionic layer where it forms a well-delimited plexus (Krettek and Price, 1977a). Thalamocortical projection fibres in contrast split up into endramifications mainly in layers IV–IIIc. The anterogenual areas receive only a sparse fibre endowment from intralaminar nuclei of the thalamus; the anterior nuclei of the thalamus do not project to this territory (Vogt et al., 1979). The anterogenual core is rather richly endowed with fibres stemming from the basal nucleus of the amygdala (Vogt et al., 1979). This rich supply with amygdalo-cortical fibres might well account for the pronouncedly internoteniate character of the anterogenual core.

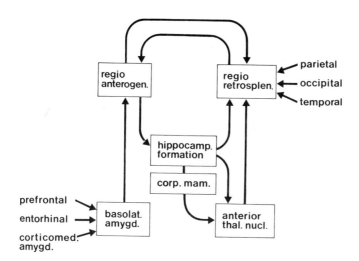

Fig. 19. Simplified scheme of the main connections of the proisocortex

Stimulation of the anterior cingulate gyrus elicits highly integrated motor responses such as stretching or sucking-in (Talairach et al., 1973), a finding which is in accordance with the magnocellular and internocrassior character of the anterogenual core.

A rather simplified scheme of some of the main connections of the cingulate proisocortical regions is given in Fig. 19.

7 The Mature Isocortex

The extension of the mature isocortex increases considerably the higher an animal is in the phylogenetic scale. In primitive forms a single structure, the lamina tecti, serves for the processing of visual, acoustic, and somatosensory information. The development of projection nuclei of the thalamus in line with that of the isocortex results in the differentiation of individual territories for each of these functions. Already in the primitive mammalian brain a visual, acoustic, and sensory-motor cortex can be distinguished (Diamond, 1967; Diamond and Hall, 1969; Casseday et al., 1976; Zilles et al., 1978).

These fundamental parts of the isocortex become gradually refined as the phylogenetic scale is ascended. Relatively late in cortical evolution, "core" fields appear which attract a particularly massive thalamic input. Hence, the structure of the elaborate human core fields differs markedly from that of the non-primate mammalian brain. The convenient practice to use the same term or number (e.g., striate area, or area 17) for designation of a certain cortical field in different species does much to obscure this fact. The core fields — or primary fields — are surrounded by "belt" areas, or secondary areas. In the final evolutionary elaboration of the isocortex progressively enlarging homotypical territories outside the core and belt areas appear, frequently denoted as "association" areas, "integration" areas, "generalized" or "uncommitted" cortex (von Bonin et al., 1942; Penfield, 1966; Diamond, 1967; Holloway, 1968; Diamond and Hall, 1969; Sanides, 1969; Sanides and Hoffmann, 1969; Brown, 1972).

Ontogenetically, the terminal areas of the large sensory projection systems as well as the primary motor field differentiate early followed by the secondary areas of the belt and comparatively late by the "association" areas (Conel, 1939; Jacobson, 1963; Yakovlev and Lecours, 1967; Holloway, 1968).

As age advances, the tertiary fields show the first signs of alteration, i.e., a decrease in overall myelin density, whereas the core fields remain well preserved for a longer time (Yakovlev and Lecours, 1967).

7.1 The Occipital Lobe (Figs. 20, 21, 22, 34)

Striate Area: Campbell (1905, Fig. 35) visuo-sensory area; Smith (1907, Fig. 36) area striata; Brodmann (1909, Fig. 37) area striata (17); von Economo and Koskinas (1925, Figs. 39, 40) area striata (OC); Bailey and von Bonin (1951, Fig. 41) isocortex koniosus striatus occipitalis; Sarkissov et al. (1955), Figs. 42, 43) field 17.

Parastriate Area: Campbell (1905, Fig. 35) part of the visuo-psychic area; Smith (1907, Fig. 36) area parastriata; Brodmann (1909, Fig. 37) and Vogt (1929) area occipitalis (18); von Economo and Koskinas (1925, Figs. 39, 40) area parastriata (OB); Bailey and von Bonin (1951, Fig. 41) part of isocortex parakoniocorticalis occipitalis; Sarkissov et al. (1955, Figs. 42, 43) field 18.

Occipital Magnopyramidal Region: Campbell (1905, Fig. 35) part of the temporal area, Smith (1907, Fig. 36) part of area temporo-occipitalis; Brodmann (1909, Fig. 37) part of area occipitotemporalis (37) and area peristriata (19); von Economo and Koskinas (1925, Figs. 39, 40) part of area parietalis basalis (PH_o) and area peristriata magnocellularis (OA_m); Bailey and von Bonin (1951, Fig. 41) part of isocortex parakoniocorticalis occipitalis; Sarkissov et al. (1955, Figs. 42, 43) lower parts of fields 19 and 37 ac.

The occipital lobe of the human brain shows clearly the aforementioned tripartition into a core, a belt, and a territory of "association" fields. Three main cortical types can be distinguished which form the striate area, the parastriate area, and the peristriate region.

7.1.1 The Striate Area

The gross morphology of the core field which receives a point-to-point projection of the visual field via the lateral geniculate body is well known (Filimonoff, 1932; von Bonin, 1942; Polyak, 1957; Whitteridge, 1973; Hendrickson et al., 1978). The abrupt appearance of the line of Gennari allows for the precice delineation of the striate area even in fresh preparations. The field is found almost entirely on the medial surface of the hemisphere (Figs. 20, 34). In subhuman primates it extends also to a varying degree on to the superolateral facies (Solnitzky and Harman, 1946; von Bonin and Bailey, 1961; Allman and Kaas, 1971b; Garey and Powell, 1971; Hubel and Wiesel, 1972, 1977; Kaas et al., 1972; Lund, 1973; Lund and Boothe, 1975; Diamond, 1976).

The Striate Area

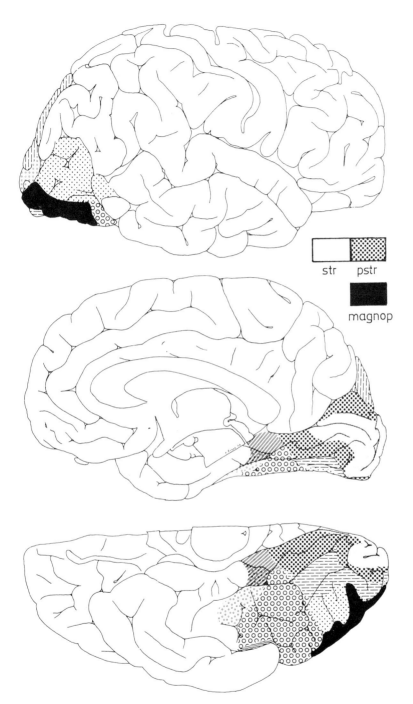

Fig. 20. Map of the pigmentoarchitectonic areas of the human occipital lobe. Note the location of area striata (*str*), area parastriata (*pstr*), and area peristriata magnopyramidalis (*magnop*). For details see H. Braak (1977)

The visual cortex of primates is remarkably narrow despite its being formed of a greater number of cellular layers than usually occur in the isocortex (Brodmann, 1904). These layers cannot readily be homologized to those found in the visual cortex of the subprimate mammalian brain (Sanides, 1972). If only descriptions of the primate striate area are taken into account, there are nonetheless remarkable differences in the nomenclature used by the various authors (Hassler and Wagner, 1965; Billings-Gagliardi et al., 1974; Valverde, 1977).

The *molecular layer* is almost devoid of nerve cells. It is narrow as compared to the corresponding laminae in the periallocortex and proisocortex. Subjacent to the pia mater there is an inextricable feltwork of expanded processes of fibrillary astrocytes forming the external glial layer (Niessing, 1936; Ramsey, 1965; Thomas, 1966; Janzen, 1967; Haug, 1971; Bondareff and McLone, 1973; Lopes and Mair, 1974; E. Braak, 1975; E. Braak et al., 1978). Just below this sheath a great number of astrocytic perikarya can be found disseminated throughout the outer third of the layer. The outer reaches of the molecular layer contain a dense accumulation of myelinated fibres. Preferential orientation of these fibres has been demonstrated in the rabbit's striate area (Fleischhauer and Laube, 1979; Maurer and Fleischhauer, 1979). Cellular processes — predominantly the terminal tufts of underlying pyramidal cells and axonal terminal ramifications — form the greater part of the layer. Only a few nerve cells are irregularly scattered throughout the layer (E. Braak, 1978b).

The *corpuscular layer* is mainly composed of small pyramidal cells with slender dendrites sparsely endowed with spines and a delicate axon passing downwards from the basis of the soma. The apical dendrites are relatively short as compared to the basal ones and frequently tilted at various angles to the cortical surface. A real apical dendrite can even be missing. The outline of the cell body gives therefore no sure criterion for distinguishing second-layer pyramids from polygonal stellate cells. The second-layer pyramidal cells are sparsely pigmented.

Fig. 21. *Upper half* Nissl and myelin preparation of the striate area of the human brain, 15 μm thick sections cut perpendicular to the cortical surface. Note the broad band of small cells which is generally referred to as layer IVc in the Nissl preparation. This picture demonstrates that a further subdivision of the broad and heterogeneously composed band is almost impossible with the aid of a Nissl preparation. The myelin preparation shows the broad stripe of Gennari (*Ge*) which is followed by a pallid zone and another myelin-rich stripe which is extremely narrow (*4d*). *Lower half* Pigment preparation of the human striate area (800 μm). Note the narrow differently pigmented laminae ($PIVc\alpha$, $PIVc\beta$, $PIVd$, PVa) which in the Nissl preparation form the broad parvocellular IVc. Layer $PIVc\beta$ is particularly conspicuous in pigment preparations and is a unique feature of the striate area

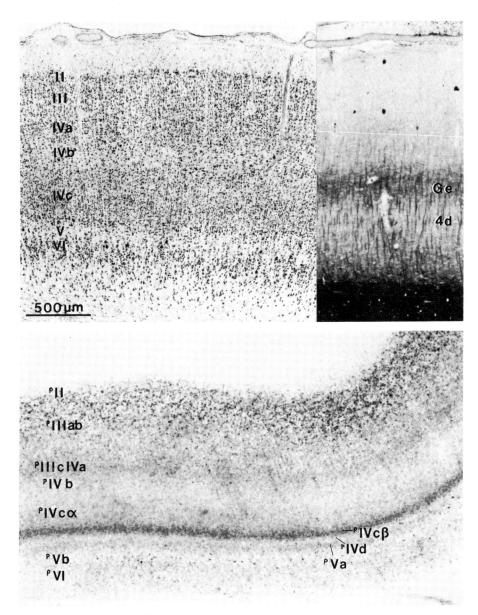

Fig. 21. Legend see p. 66

The outer parts of the *pyramidal layer* harbour characteristically formed and modestly pigmented pyramids increasing in size from above downwards. Numerous side branches arise from the shaft of the apical dendrites. The cell body issues a great number of richly branching basal dendrites. Both types of dendrites are abundantly provided with spines. The spine-free proximal zones are extended. These features all together indicate a high level of differentiation.

Laminae II and IIIab contain many short-axoned stellate cells. Those with radially oriented cell bodies (double-bush cells: Ramón y Cajal, 1909) lack pigment deposits. The pigment-laden stellate cells show small polygonal cell bodies issuing only a few smoothly contoured dendrites. Their packing density is highest in the lower reaches of the corpuscular layer but they also penetrate widely into the pyramidal layer.

Radially oriented bundles of myelinated fibres extend up to the lower reaches of IIIab.

A sudden increase in the nerve cell packing density indicates the border of layer IIIab to *layer IIIc–IVa*, which is mainly composed of an admixture of characteristically formed third-layer pyramids and spiny "stellate" cells which are typical components of the granular layer in the striate area (Garey, 1971; Le Vay, 1973; Lund, 1973; Jones, 1975). The layer is therefore designated IIIc–IVa. The spiny "stellate" cells can be considered a conspicuous variety of modified pyramids with more or less horizontally aligned spindle-shaped cell bodies (see Chap. 2.2). A few infrequently branching dendrites arise from opposite poles of the soma. They are sparsely invested with spines. The axon heads downwards in a straight line. At a short distance from the soma some stout collaterals are given off. The cell bodies are almost barren of pigment deposits.

Besides these cells, many short-axon stellate cells with smoothly contoured dendrites are scattered throughout the layer. Often, not only the dendrites but also the axonal ramifications display a beaded appearance. The cell bodies are devoid of pigment. Without remarkable variations they can also be found in great numbers in the following laminae.

Layer IVb is distinguished by its low packing density of nerve cells. The dominant cell type is a large variant of the spiny "stellate" cells, the "solitary" cell of Ramón y Cajal (1900a, 1909). Far-reaching dendrites arise from their horn-shaped cell bodies running horizontally or slightly upwards. The dendrites are richly studded with spines. The axon emerges from a downwards-directed hillock and descends in a straight line to enter the white matter. The cells are almost devoid of pigment.

The layer is filled with a dense plexus of myelinated fibres which form the line of Gennari, a hallmark of the striate area of primates. The line shows blurred upper and lower borders and extends partly into the layer IVc. As to its location and extension it corresponds to the external tenia (Figs. 21, 22).

The upper border of *lamina IVc* is indicated by an abrupt increase in the packing density of the nerve cells and a sharp decrease in their average size. The layer is split into an upper part, IVcα, which is mainly composed of sparsely pigmented spiny "stellate" cells and a lower part, IVcβ, the most significant cell type of which is a small modified pyramidal cell which gives off a few short dendrites modestly invested with spines. The axon initially heads downwards but bends back at a short distance from the soma to ramify in the superficial layers (Lund, 1973; Lund and Boothe, 1975; H. Braak, 1976b; E. Braak, 1978a,b). The only small envelope of cytoplasm around the large nucleus contains some large lipofuscin granules which are composed of intensely stained parts and light vacuoles. Coarse pigment granules of this type can only infrequently be found in cortical cells. It seems worthwhile to point out that the same type of pigment occurs in the main cell type of the presubicular parvopyramidal layer (see Chap. 5.2.1).

Delineation of IVcβ is difficult in Nissl preparations since it is embedded within a cell-rich stripe comprising the layers IVcα, IVd, and Va as well (Fig. 21). Pigment preparations by contrast reveal the clear-cut borderlines of IVcβ (Fig. 21). Like the line of Gennari, layer IVcβ is a unique feature of the striate area.

The dominant constituents of IVcβ are known to receive the bulk of projection fibres generated from the parvocellular lamellae of the lateral geniculate body (Hubel and Wiesel, 1972; Lund, 1973).

A narrow zone which is prevalently filled with non-pigmented small pyramidal cells (IVd) follows. Myelin preparations cut exactly perpendicular to the cortical surface display a tenuous line of horizontally adjusted fibres within the limits of this layer (Fig. 21).

Layer Va is also remarkably attenuated and consists of numerous tiny pyramidal cells. When surpassing the boundary of the striate area, the layer broadens abruptly and fuses with the ganglionic layer of the parastriate fields. The remarkable attenuation of ^{P}Va is reminiscent of that found in the retrosplenial fields (see Chap. 6.1). The apical dendrite of the small pyramids is particularly thin and bears only a few spines. It does not extend up to the molecular layer. The slender cell bodies contain finely grained and only modestly tinged pigment.

The cell-sparse *lamina Vb* is broad and contains medium-sized to large smoothly contoured stellate cells and medium-sized pyramids. Now and then pyramidal cells occur which are distinguished by their enormous size. These Meynert pyramids generate a thin apical dendrite and a great number of far-reaching basal dendrites, one of which is a particularly extended process (Clark LeGros, 1942; Shkol'nik-Yarros, 1971; Chan-Palay et al., 1974; Palay, 1978; Braitenberg and Braitenberg, 1979). In the human brain, the dendrites of mature Meynert pyramids lack spines or bear only a few (H. Braak, 1976b; E. Braak, 1978b). The axon contributes to the white matter. The perikaryon is almost barren of lipofuscin granules. Large Nissl bodies are distinctly evident.

Layer Vb exhibits loosely distributed myelinated fibres which form the inner line of Baillarger. On the average, the striate area is rich in fibres and therefore the inner line is seen as through a veil. Suitable differentiated myelin preparations nevertheless display the line (Vogt and Vogt, 1919; Kawata, 1927; Kirsche and Kirsche, 1962), a fact which probably entails revision of the claim that the striate area shows the singulostriate characteristic (Sanides and Gräfin Vitzthum, 1965a; Gräfin Vitzthum and Sanides, 1966; Sanides, 1972). The euradiate field seems rather to be of the bistriate and extremely externodensior type.

Pigment preparations display a clear-cut but narrow internal tenia which gives the field an extremely externoteniate character.

The *multiform layer* is split into a narrow densely pigmented upper part (^{P}VIa) and a broad modestly pigmented lower one (^{P}VIb). Its main constituents are pyramidal cells issuing slender apical dendrites. They are heavily invested with spines as they pierce lamina V but become almost barren of spines in the laminae IVcβ and IVcα (Lund, 1973).

7.1.2 The Parastriate Area

The visual core field is throughout its circumference accompanied by the parastriate area. The spread of the belt area is again almost restricted to the medial facies of the hemisphere. Its anterior extremity is concealed in the common trunk of the parieto-occipital and the calcarine sulcus (Figs. 20, 34).

The boundary between the core and the belt area is sharply drawn. Most of the layers of the parastriate field show alterations as compared to

---

Fig. 22. *Left upper parts* Border zone between the area striata and area parastriata of man. Sections successively cut from the same block of tissue for Nissl and myelin staining (15 μm). Note the shallow groove of the cortical surface in the vicinity of the border. The myelin preparation displays the subconjunctostriate characteristic of the parastriate area and within the striate area besides the stripe of Gennari (*G*) the bridging stripe of Baillarger (*B*). *Lower part* Pigment preparation (800 μm) of the border zone. Note the tripartition of the external tenia (^{PI}Va, ^{PI}Vb, ^{PI}Vc) in the parastriate field. *Right upper part* Area peristriata magnopyramidalis. Pigment preparation (800 μm) cut perpendicular to the cortical surface. Note the conspicuous accumulation of large pigment-laden IIIc-pyramids (*arrow*)

The Parastriate Area

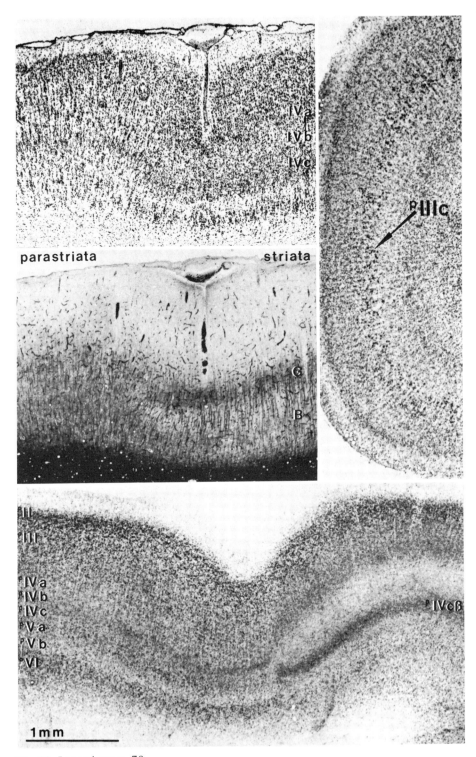

Fig. 22. Legend see p. 70

the corresponding laminae of the striate area; only the *superficial laminae* (II and IIIab) remain much the same (Valverde, 1978). The upper reaches of the pyramidal layer display a fair number of stout horizontally adjusted myelinated fibres. These form the line of Kaes-Bechterew which gives the visual belt area an extremostriate character.

The lower reaches of the *pyramidal layer* (IIIc) are mainly filled with medium-sized to large pyramids. Myelin preparations display thick radially oriented fibre bundles vanishing at about the upper border of IIIc. The existence of a real sublayer IIIc distinguishes the visual belt area from the core field. The external tenia penetrates into the lower parts of the pyramidal layer. The large IIIc-pyramids remain therefore almost devoid of pigment. The upper parts of IIIc nevertheless show typically well-pigmented IIIc-pyramids. A weak radially oriented striation is recognizable in these parts.

The *granular layer* contains a wealth of densely packed small nerve cells. Though the individual cells are poorly invested with pigment granules, their great numbers result in the appearance of a weakly tinged stripe located in the middle of the external tenia. The stripe is in continuation with sublamina IVcβ of the striate area. Tripartition of the external tenia (^{P}IVa, ^{P}IVb, ^{P}IVc) is a distinguishing characteristic of the parastriate area and does not elsewhere occur in the isocortex (Fig. 22). The pigment preparation therefore permits also clear delineation of the peripheral border of the belt towards the peristriate region (H. Braak, 1977). This border often defies recognition in Nissl or myelin stains (von Economo and Koskinas, 1925; Bailey and von Bonin, 1951).

The *ganglionic layer* is divisible into a cell-rich Va harbouring many weakly pigmented pyramids and a cell-sparse Vb which corresponds to the internal tenia. The outstanding width of the external tenia as compared to the narrowness of the internal one gives the parastriate field a markedly externoteniate character. The lower reaches of the pyramidal layer, the granular layer, and in particular the ganglionic layer contain a wealth of myelinated horizontal fibres. The intrastriate zone (5a) is also richly endowed with fine "ground" fibres, which results in the conjunctostriate type of the parastriate field. The average myelin content is high (typus dives).

The *multiform layer* is accentuated in pigment preparations. The staining capacity of the pigment deposits in the upper parts of ^{P}VI is markedly higher than in ^{P}III or ^{P}V or than in the multiform layer of the striate area. The cell density is highest at the top, decreasing gradually as the layer is descended.

The Parastriate Border (Limes parastriatus). Close to the core field the parastriate area shows special features which allow for delineation of a border zone (Fig. 22). Functionally this territory serves to connect neighbouring points of the visual field on both sides of the vertical meridian (Vogt, 1929; Myers, 1962, 1965; Sanides and Gräfin Vitzthum, 1965b; Zeki, 1970; Glickstein and Whitteridge, 1974; Shoumura et al., 1975).

A limited number of large pyramids crop up within the lower reaches of the third layer forming the "limes parastriatus gigantopyramidalis" (OBγ: von Economo and Koskinas, 1925). In the immediate vicinity of the border, the radiate fibre bundles turn out to be of particular stoutness (limiting bundles: Sanides and Gräfin Vitzthum, 1965b). In addition, the inner line of Baillarger can easily be followed for a certain distance into the striate area before becoming less clearly recognizable (Fig. 22, bridging stripe of Baillarger: Sanides and Gräfin Vitzthum, 1965b).

7.1.3 The Peristriate Region

The visual belt area is surrounded by the peristriate region which is comprised of many cortical areas which have some traits in common. They spread over the remaining parts of the occipital lobe, but extend also widely into the basal parts of the temporal lobe (Figs. 20, 34).

A detailed description of the individual peristriate areas is outside the scope of the present text. The reader is therefore referred to the literature on the subject (Filimonoff, 1932; Beck, 1934; Lungwitz, 1937; von Bonin et al., 1942; Zeki, 1969; Whitteridge, 1973; H. Braak, 1977). In this context it seems advisable to concentrate on only the magnopyramidal peristriate area which is morphologically an outstanding part of the peristriate region (H. Braak, 1977).

The *area peristriata magnopyramidalis* covers the edge of the inferior facies and the superolateral facies of the occipital lobe not far behind the preoccipital incisure (Figs. 20, 34).

The superficial layers (II and IIIab) are rich in sparsely pigmented pyramids and small pigment-laden stellate cells. The line of Kaes-Bechterew is absent. Pigment preparations display a broadening of the third layer which occurs at the expense of the external tenia. ^{P}IIIc shows a clear striation. Besides a tightly packed population of common third-layer pyramids there occur remarkably large pyramids which store a great amount of pigment in an aggregate. These pigment-laden IIIc-pyramids are irregularly dotted about the lower reaches of the third layer and subjacent parts of the external tenia. The cells keep far apart from each other (Fig. 22).

The breadth of the external tenia matches roughly that of the internal one. As is the case in all the fields of the peristriate region, the external tenia does not contain a line which is comparable to that found in the parastriate field (^{P}IVb). Myeloarchitecturally, the field is of the subconjunctostriate type. The lower main stratum (V–VI) does not differ markedly from that of the parastriate field.

Summing up, the analysis of the occipital lobe reveals a heterotypical *core field* which is parvocellular and hypergranular, rich in myelinated fibres (typus dives) and poor in overall pigmentation (typus clarus). It shows a markedly externocrassior, externodensior, and externoteniate characteristic. All these features can be considered hallmarks of a sensory core field in the mature isocortex.

The *belt area* still shows a prominent granular layer but the layer does not predominate. The belt area is rich in myelinated fibres, its average pigmentation is of medium density. It is externocrassior and markedly externoteniate. Since the field is of the subconjunctostriate type, the possible preponderance of one of the lines of Baillarger cannot be evaluated.

The *"association" areas* of the peristriate region display a homotypical cortex with an average content of both myelinated fibres and pigment. The peristriate region is for the most part equocrassus. Its subconjunctostriate characteristic is not especially conducive to permit evaluation of the lines of Baillarger. Both teniae are approximately equally wide.

The extended territories outside the core and the belt generally reveal the characteristics of an unspecialized calibrated cortex without marked predominance of either large pyramids or small "granule" cells or prevalence of one of the lines of Baillarger or one of the teniae.

7.2 The Temporal Lobe (Figs. 23, 34)

Temporal Granulous Core: Campbell (1905, Fig. 35) audito-sensory area; Smith (1907, Fig. 36) part of "Heschl's gyri"; Brodmann (1909, Fig. 37) part of area temporalis transversa externa (42); von Economo and Koskinas (1925, Figs. 39, 40) area supratemporalis granulosa (TC); Bailey and von Bonin (1951, Fig. 41) isocortex koniosus supratemporalis.

Fig. 23. Map of the pigmentoarchitectonic areas of the human temporal lobe. Note the ▶ location of area temporalis granulosa (*gran*), area temporalis progranulosa (*progr*), area temporalis paragranulosa (*paragr*), area temporalis magnopyramidalis centralis (*magn. py.c*), area temporalis magna (*magna*), and area temporalis stratiformis (*strat*). For details see H. Braak (1978c)

The Temporal Lobe

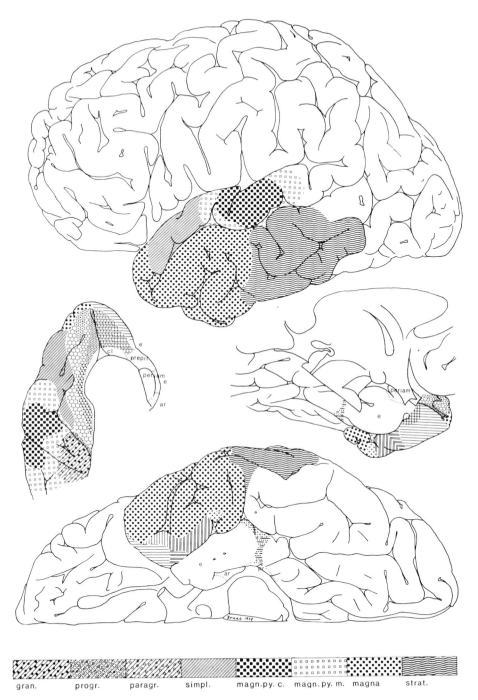

Fig. 23. Legend see p. 74

Temporal Paragranulous Belt: Campbell (1905, Fig. 35) posterior parts of the audito-psychic area; Smith (1907, Fig. 36) part of area temporalis superior; Brodmann (1909, Fig. 37) medial parts of area temporalis superior (22); von Economo and Koskinas (1925, Figs. 39, 40) parts of area supratemporalis magnocellularis (TB); Bailey and von Bonin (1951, Fig. 41) part of isocortex parakoniocorticalis supratemporalis; Sarkissov et al. (1955, Figs. 42, 43) part of field 22.

Temporal Magnopyramidal Region: Campbell (1905, Fig. 35) posterior parts of the audito-psychic area; Smith (1907, Fig. 36) part of area temporalis superior; Brodmann (1909, Fig. 37) posterior and lateral parts of the area temporalis superior (22); von Economo and Koskinas (1925, Figs. 39, 40) parts of area temporalis superior posterior (TA_1) and area supratemporalis magnocellularis (TB); Bailey and von Bonin (1951, Fig. 41) part of isocortex parakoniocorticalis supratemporalis, Sarkissov et al. (1955, Figs. 42, 43) part of field 22.

Temporal Uniteniate Region: Campbell (1905, Fig. 35) lower parts of the temporal area; Smith (1907, Fig. 36) posterior parts of area temporalis media et inferior; Brodmann (1909, Fig. 37) posterior parts of area temporalis media et inferior (21 et 20); von Economo and Koskinas (1925, Figs. 39, 40) posterior parts of area temporalis media (TE_1) and area temporalis inferior (TE_2); Bailey and von Bonin (1951, Fig. 41) part of isocortex eulaminatus temporalis inferior; Sarkissov et al. (1955, Figs. 42, 43) parts of fields 21 and 37.

The temporal lobe can be subdivided into three major territories: the allocortical zone, which is particularly expanded anteromedially, and two isocortical territories, one of which spreads over the superior temporal gyrus and differs markedly from that covering the middle and inferior temporal convolutions.

The isocortex of the superior temporal gyrus is comprised of a granulous core surrounded by belt areas and by another ring of "association" area.

7.2.1 The Temporal Granulous Core

The area temporalis granulosa occupies only a small part of the first transverse gyrus of Heschl. The field is totally buried in the depth of the lateral sulcus and situated close to the insula (Fig. 26). Its location may vary to some extent from individual to individual. The lateral border is often marked by a shallow groove. Frequently, the core field spreads over only a posterior portion of the transverse gyrus and covers to a variable extent parts of the temporal plane. The long axis of the elongated stretch of coniocortex generally runs in parallel with the transverse gyrus. The average extension of the core is astoundingly small (Flechsig, 1908; Pfeifer, 1920, 1936; von Economo and Koskinas, 1925; Rose, 1935).

The location of the acoustic core in primates has repeatedly been documented experimentally (Ades and Felder, 1942; Hind et al., 1958; Woolsey, 1961; Massopust et al., 1968; Sanides, 1972; Mesulam and Pandya, 1973; Pandya and Sanides, 1973; Forbes and Moskowitz, 1974; Casseday et al., 1976; Imig et al., 1977; Oliver and Hall, 1978).

There is much evidence from anatomical and physiological investigations that the granulous core reveives the bulk of thalamo-cortical fibres from the acoustic radiation (Locke, 1961; Mesulam and Pandya, 1973; Harrison and Howe, 1974; Jones and Burton, 1976) and serves as a tonotopically organized primary auditory field (Merzenich and Brugge, 1973; Merzenich et al., 1973).

The cortex is unusually thick as compared to granulous core fields in other parts of the brain. It is particularly conspicuous in its maximum density of myelinated fibres (Hopf, 1954, 1968) and in its pallid appearance in pigment preparations (H. Braak, 1978b,c) (Figs. 24, 25).

The *molecular layer* contains a small number of horizontal cells of Cajal. Its superficial parts are rich in tangentially adjusted myelinated fibres.

The *corpuscular layer* is rather wide and comprises a wealth of tiny pyramids and pigment-laden stellate cells.

The *pyramidal layer* as well is mainly composed of tightly packed small pyramids, the size of which is not subject to a marked increase as the deep border of the layer is approached. The small pyramids are perpendicularly arranged in thin lines (rainshower formation: von Economo and Koskinas, 1925). Those which are situated in the upper parts of the layer still show a discernible degree of pigmentation whereas those accommodated in the lower parts are barren of pigment. This results in a particularly blurred upper border of the outer tenia (Fig. 24). Typically, the third-layer pyramids become more strongly pigmented the deeper their position is; the reverse pattern of pigmentation found here is a distinguishing characteristic of core fields in other parts of the brain as well (retrosplenial, somatosensory, and somatomotor cortex). A weak and fine striation is recognizable throughout the pyramidal layer. Here and there some isolated pigment-laden IIIc-pyramids are dotted about the lower reaches of the third layer.

The outstandingly broad *granular layer* is customarily formed of densely packed minute pyramidal cells. They are aligned perpendicular to the cortical surface. The granular layer and suprajacent parts of the pyramidal layer as well are filled with myelinated horizontal fibres.

The *ganglionic layer* is split into an upper cell-rich zone and a cell-sparse lower one both formed of small-sized pyramids arranged in columns. The internal tenia is narrow and shows clear-cut borders. Myelin preparations display a fibre-rich intrastriate zone and a dense inner line of Baillarger.

The *multiform layer* is also filled with myelinated fibres. The packing density of its richly pigmented nerve cells is highest at the top of the layer, decreasing gradually downwards. The layer fades into the white matter without any gradation to permit separation in sublayers.

The fringe areas surrounding the core are composed of a proisocortical field anteriorly and a paragranulous belt area posteriorly (Sanides, 1972, 1975).

7.2.2 The Temporal Progranulous Field

The area temporalis progranulosa is inserted between the insular proisocortex on the one side and the parvocellular core field on the other. The borderline between the core field and the proconiocortex is abrupt. The sudden appearance of sublayer P_{IIIc} accounts for the marked attenuation of the external tenia in the progranulous field (Fig. 24).

7.2.3 The Temporal Paragranulous Belt

The area temporalis paragranulosa forms a hook-like border zone around the core (Fig. 26). Anteriorly, it abuts on the proconiocortex. In pigment preparations the pallor of the pyramidal layer distinguishes it from the progranulous field. The external tenia is clearly a step less broad than in the core field (Fig. 24).

Fig. 24. Cortex covering proximal parts of the first transverse gyrus of Heschl. Pigment ▶ preparation (800 µm) cut at right angles to the gyrus. The extension of the area temporalis granulosa is indicated (*gran*) by an abrupt widening of the external tenia (*Te*). The boundaries are marked by *triangles*. Note the appearance of sublayer P_{IIIc} in the adjoining area temporalis paragranulosa (*paragran*) and area temporalis progranulosa (*progran*)

The Temporal Paragranulous Belt

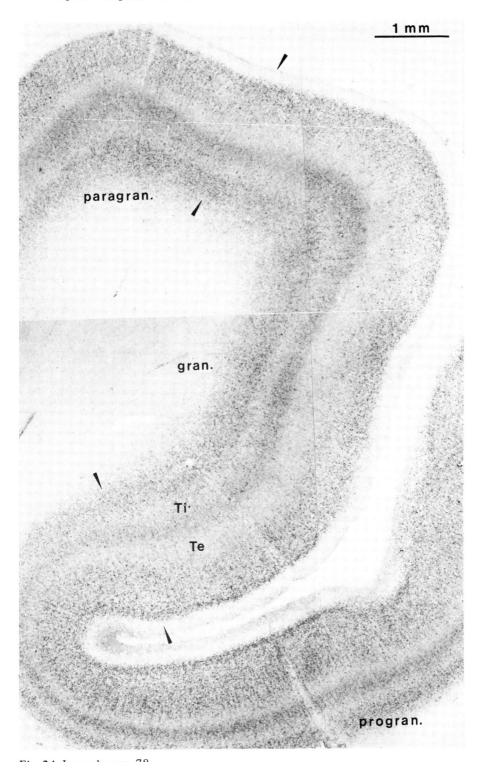

Fig. 24. Legend see p. 78

7.2.4 The Temporal Magnopyramidal Region

Proceeding lateral-wards, there follows an extended territory of equoteniate "association" areas which spread over large parts of the temporal plane, i.e., the upper bank of the first temporal convolution behind the transverse gyrus of Heschl (Fig. 26). These areas display a striking population of large pigment-laden IIIc-pyramids, thereby constituting the temporal magnopyramidal region.

Relatively small nerve cells are still the predominant constituents of the cortex. The myelin content is considerably decreased as compared to the core and the belt. The region separates into a refined central field and fringe areas mediating to the surrounding cortex (Fig. 26). The features of the central field disappear in definite stages, whereas those of the borderlands become more and more distinct. A sketch of the central field will be sufficient for an introduction into the morphology of the temporal magnopyramidal region (Fig. 25).

The Central Magnopyramidal Field. The *molecular layer* is poor in myelinated fibres.

The broad *corpuscular layer* is composed of tiny pyramids which preferably fill up the upper parts of the layer and of small stellate cells. The great number of pigment-laden stellate cells gives the lower reaches of the layer a band-like appearance.

Medium-sized and tightly packed pyramids prevail in the *pyramidal layer.* They are aligned in columns. This common type of pyramidal cell forms the inconspicuous background to the unique population of large third-layer pyramids with voluminous rounded pigment agglomerations. The pigment-laden pyramids are irregularly scattered throughout both the lower reaches of the pyramidal layer and the external tenia (Fig. 25).
A more or less abruptly occurring numerical decrease of pigment-laden IIIc-pyramids indicates the hazy transition from the central field to the fringe areas.

Fig. 25. *Upper half* Area temporalis granulosa of man. Nissl and myelin preparations successively cut from the same block of tissue (15 μm). Note the parvocellular character of the cortex and the arrangement of the nerve cells in thin lines (rain-shower formation). *Be* outer stripe of Baillarger. *Bi* inner stripe of Baillarger. *Lower half* Area temporalis magnopyramidalis centralis. Pigment preparation (800 μm). The section is cut at right angles to the first transverse gyrus and lies close to the free surface of the temporal lobe. Note the conspicuous population of large and pigment-laden IIIc-pyramids. *Inset* Area temporalis stratiformis. Pigment preparation (800 μm). Besides the external tenia (*Te*) the cortex shows a particularly narrow and sharply drawn pallid line within the lower reaches of the pyramidal layer (*arrowhead*)

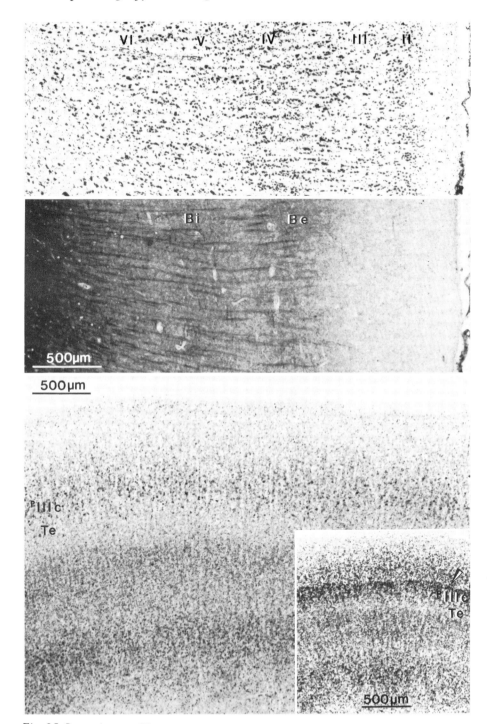

Fig. 25. Legend see p. 80

The *granular layer* is rich in minute pyramidal cells perpendicularly aligned in thin columns (organ-pipe-formation: von Economo and Koskinas, 1925). The outer Baillarger is very definite. The pallid external tenia is broad and strikes one immediately though both its upper and lower borders fail to stand out clearly.

The *ganglionic layer* is remarkably broad and harbours well-formed pyramids. It is divided into a weakly pigmented upper part and a cell-sparse lower zone, each of roughly the same width. Sublayer Va displays a fine radiate striation. A fair number of faintly pigmented pyramids can be encountered throughout the limits of the internal tenia, accounting for its less pallid appearance as compared to the outer tenia. The intrastriate zone is filled with a fair number of myelinated fibres, thereby impeding delineation of the inner line of Baillarger. Sufficiently differentiated myelin preparations display both lines equally broad and dense.

The *multiform layer* is accentuated with a clear-cut upper border. Besides sparsely pigmented constituents it harbours nerve cells stuffed with lipofuscin granules. These are most densely placed in the upper reaches and disappear gradually as the layer fuses with the white substance.

The Speech Centre of Wernicke. Location and extension of the magnopyramidal region are subjected to considerable variation not only between individuals but also between both sides of the brain. Neither the macroscopical borders of the temporal plane nor the limiting sulci allow for a simple delineation of the region by mere examination of the surface relief (H. Braak, 1978b).

The variability in the extension of cortical fields seems to increase with phylogenetic advance (Filimonoff, 1932, 1933). Until now, convincing attempts for correlation of the cortical extension of the various areas with the peculiar abilities or incapacities of the human being are missing (Vogt, 1941; Vogt and Vogt, 1940, 1942, 1954; Whitaker and Selnes, 1976; Meyer, 1977).

Areas distinguished by pigment-laden IIIc-pyramids have not as yet been recognized in the brain of subhuman primates. The posterior parts of the first temporal convolution have long been considered to represent the sensory speech centre of Wernicke, a location derived from the results of stimulation during surgery and from examinations of brains from aphasic patients afflicted with cirumscribed brain damage (Wernicke, 1874; Beck, 1936; Hopf, 1957; Penfield and Roberts, 1959; Russel and Espir, 1961; Brown, 1972; Geschwind, 1974; Celesia, 1976; Ingvar, 1976; Geschwind et al., 1979). On account of both its unique endowment with specialized pyramids which elsewhere do not occur in the temporal lobe and its peculiar location, it seems tempting to consider the temporal magnopyramidal

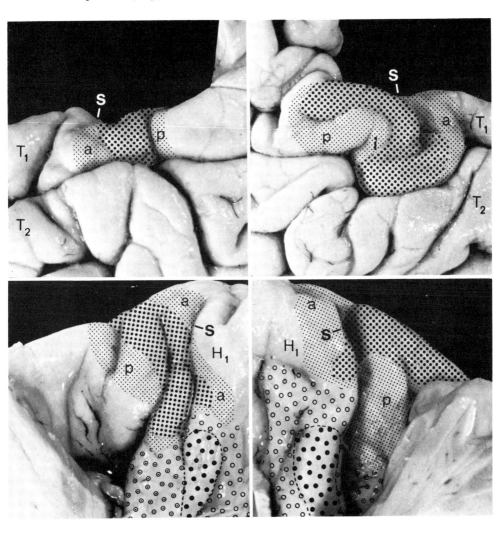

Fig. 26. Map of the acoustic core, its belt, and the temporal magnopyramidal region of man. The *upper half* shows the superolateral facies of posterior portions of the left and right temporal lobes after removal of the inferior parietal lobules. The *lower half* displays the corresponding superior facies of the left and right temporal lobes. In this brain the left temporal plane is considerably larger than the right. Only a part of the temporal magnopyramidal region spreads over the temporal plane. The magnopyramidal region appears as a lateral expansion of the paragranulous belt area. The location of the central field is marked by *thick dots*, whereas that of the marginal fields (*a* anterior, *i* inferior, *p* posterior) is given by *fine dots*. The location of area temporalis granulosa is indicated by *particularly thick dots*, that of the paragranulous belt by *circles with a core*, and that of the progranulous field by *circles*. *H1* first transverse gyrus of Heschl, *S* sulcus temporalis profundus, T_1 and T_2 superior and middle temporal gyrus

region as forming an essential part of the sensory speech centre of Wernicke (H. Braak, 1978b).

Briefly put, the structural analysis of the isocortex covering the superior temporal convolution discloses a parvocellular *core field* which is particularly rich in myelinated fibres (typus dives) and remarkably poor in pigment (typus clarus). The field is externocrassior and extremely externoteniate. The inner line of Baillarger prevails.

The surrounding paragranulous *belt area* is still granularized, though not to such an extent as shown by the core field. It is also strongly invested with myelinated fibres and shows a pallid appearance in pigment preparations. The externoteniate characteristic is less expressed than in the core. Again the inner line of Baillarger is predominant.

The *"association" areas* (magnopyramidal region) can already be classified with the homotypical cortex. There is only an average supply with myelinated fibres. The lines of Baillarger are equally dense and wide. The overall pigmentation is neither weak nor dense. The cortex is equoteniate.

This sequence of leading features characterizing the core, the belt, and "association" areas is reminiscent of that found in the occipital lobe.

7.2.5 The Temporal Uniteniate Region

The lightly pigmented biteniate areas spreading over the superior temporal gyrus differ clearly from the strongly pigmented uniteniate fields covering the middle and inferior temporal convolutions.

Within the latter territory two expanded fields, the area temporalis magna and the area temporalis stratiformis, can be distinguished (H. Braak, 1978c). Both areas have many features in common. The expanded stratiform area forms a link between the temporal isocortex and the peristriate areas (Fig. 23). Experimental investigations in the brains of primates provide data buttressing the conjecture that this stretch of cortex might be concerned with visual discrimination (Chow, 1961; Zeki, 1969, 1977; Iversen, 1970; Jones and Powell, 1970e; Allman and Kaas, 1971a, 1974; Mishkin, 1972; Dean and Cowey, 1977).

The *area temporalis magna* shows a broad pyramidal layer. Its lower zone rich in pigment (P_{IIIc}) appears enlarged at the expense of the pallid P_{IIIab} which often is so tenuous as to defy recognition. The lower border of P_{IIIc} is sharply drawn. The granular layer is narrow. The outer tenia appears also as an attenuated line. The broad ganglionic layer is dominated by feebly pigmented pyramids showing a gradual decrease in their packing density as the multiform layer is approached. The inner tenia is lacking.

The *area temporalis stratiformis* displays roughly the same features as the aforementioned field. Its distinguishing characteristic is a sharply outlined pallid stripe situated in the lower reaches of the pyramidal layer (Fig. 25, arrow-head). Its breadth ranges from 40–60 μm. The line is empty of pigmented nerve cells and takes a slightly wavy course parallel to the upper border of the strongly pigmented P_{IIIc}. In Nissl and myelin preparations the line is not displayed to advantage. Combined Nissl pigment preparations reveal that the line is not a result of densely packed non-pigmented nerve cell bodies or glial cells. Within the limits of the line there is also no unusual vascular supply or a dense plexus of myelinated fibres (H. Braak, 1978c).

7.3 The Parietal Lobe (Fig. 34)

Parietal Granulous Core: Campbell (1905, Fig. 35) anterior parts of the postcentral area; Smith (1907, Fig. 36) anterior parts of the area postcentralis A; Brodmann (1909, Fig. 37) area postcentralis oralis (part 3b); Vogt (1911, Fig. 38), M. Vogt (1928), Gerhardt (1940), Batsch (1956), Gräfin Vitzthum (1959), Hopf (1969/70, 1970): area grossofibrosa paradoxa (69); von Economo and Koskinas (1925, Figs. 39, 40) area postcentralis oralis granulosa (PB_1), and area postcentralis oralis simplex (PB_2); Bailey and von Bonin (1951, Fig. 41) isocortex koniosus postcentralis; Sarkissov et al. (1955, Figs. 42, 43) part of field 3.

Parietal Paragranulous Belt: Campbell (1905, Fig. 35) posterior parts of the postcentral area; Smith (1907, Fig. 36) posterior parts of the area postcentralis A; Brodmann (1909, Fig. 37) anterior parts of the area postcentralis intermedia (1); Vogt (1911, Fig. 38) anterior parts of area grossofibrosa aequidensa (70); von Economo and Koskinas (1925, Figs. 39, 40) part of area postcentralis intermedia (PC); Bailey and von Bonin (1951, Fig. 41) part of isocortex parakoniocortalis postcentralis; Sarkissov et al. (1955, Figs. 42, 43) anterior parts of field 1.

Parietal Magnopyramidal Region: Campbell (1905, Fig. 35) upper parts of the temporal area and anteriorly adjoining parts of the parietal area and the intermediate postcentral area; Smith (1907, Fig. 36) area parietalis inferior B and C plus adjacent parts of area postcentralis B; Brodmann (1909, Fig. 37) area supramarginalis (40); Vogt (1911, Fig. 38) area propeastriata dives et pauper (88 and 89); von Economo and Koskinas (1925, Figs. 39, 40) part of area supramarginalis (PF); Bailey and von Bonin

(1951, Fig. 41) part of isocortex eulaminatus parietalis; Sarkissov et al. (1955, Figs. 42, 43) inferior parts of fields 1, 2, and 40.

The parietal lobe accomodates the somatosensory projection cortex which lies along the anterior wall of the postcentral gyrus. It is accompanied by well-developed belt areas and particularly expanded "association" fields.

7.3.1 The Parietal Granulous Core

The parvocellular core is totally buried in the depth of the central sulcus. In general it does not extend on to the medial facies of the hemisphere. Superolaterally, it can only be followed down to the level of the inferior frontal sulcus where it abuts upon the subcentral region (Fig. 34).

The location of the somatosensory core in primates has repeatedly been demonstrated (Powell and Mountcastle, 1959a,b; Sanides and Krishnamurti, 1967; Sanides, 1968; Jones and Powell, 1970d; Pubols and Pubols, 1971; Jones et al., 1978; Merzenich et al., 1978; Whitsel et al., 1978).

There is much evidence that the postcentral core receives the bulk of thalamo-cortical fibres forming the somatosensory radiation (Moffie, 1949; Clark, Le Gros and Powell, 1953; Chow and Pribram, 1956; Jones and Powell, 1970d).

The precise somatotopical arrangement within the projection is reflected by a corresponding representation of the body scheme in the postcentral core and belt (Penfield and Boldrey, 1937; Woolsey and Fairman, 1946; Penfield and Rasmussen, 1950; Werner and Whitsel, 1971; Roland and Larsen, 1976; Kaas et al., 1979).

The cortex of the core is signally narrow. Thinnest at the upper margin of the hemisphere, the cortex broadens gradually as one approaches the inferior extremity of the parietal core field.

The *molecular layer* is poor in cells and of medium breadth. Its upper parts contain a wealth of myelinated fibres.

The *corpuscular layer* is richly stocked with small pyramids and pigment-laden stellate cells.

The *pyramidal layer* is prevalently formed of small tightly packed pyramids of approximately uniform size. Their lipofuscin deposits diminish from above downwards. The outer tenia is therefore extremely broad and displays a blurred upper boundary. Besides the granular layer the external tenia covers wide parts of the pyramidal layer as well.

The *granular layer* impresses itself by its great number of densely packed small pyramids and stellate cells which partly tend to penetrate

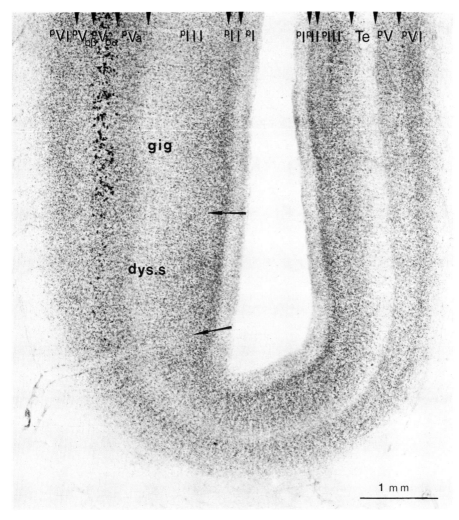

Fig. 27. Pigment preparation cut at right angles to the central sulcus close to the paracentral lobule. The section displays the gigantoganglionic core field (*gig*), the superior dysganglionic area (*dys.s*), and the following parietal granulous core which is markedly externoteniate. *Arrows* indicate the boundaries of the fields

into the pyramidal layer (Fig. 28). In pigment preparations the layer is split into an upper pallid part and a lower one which accommodates perikarya with a few feebly tinged lipofuscin granules. The latter part resembles to a certain extent similarly pigmented laminae in other coniocortices such as $^pIVc\beta$ in the striate area, ^pIVb in the retrosplenial core, or the parvopyramidal layer in the presubiculum.

The *ganglionic layer* consists of a tenuous zone which is formed of weakly pigmented ^pVa-pyramids and a narrow internal tenia (^pVb).

The layer is filled with myelinated fibres which are especially crowded in its lower parts. The inner line of Baillarger stands out in myelin preparations (Fig. 28).

The narrow *multiform layer* consists of almost uniformly pigmented nerve cells. The deep border towards the white substance is sharply drawn (Fig. 28).

7.3.2 The Parietal Paragranulous Belt

The belt extends in a strip-like fashion posterior and parallel to the core field. Starting out at the anterior wall of the postcentral gyrus it stretches out on to the free surface of the brain. It reaches its greatest expansion close to the upper margin of the hemisphere, narrowing down and receding into the central sulcus as its inferior extremity is approached. The belt ends at about the latitude of the inferior frontal sulcus (Figs. 28, 34).

The *molecular layer* calls for no special remark.

The *corpuscular layer* retains its characteristics as a cell-rich band.

The *pyramidal layer* is of medium breadth and contains, in contrast with the coniocortex, pyramidal cells growing larger the deeper their position is. Also the amount of pigment increases considerably as the layer is descended, a fact which results in a clear-cut lower border of P_{IIIc}. A fair number of myelinated fibres in the superficial parts of the layer form the line of Kaes-Bechterew (Batsch, 1956).

The *granular layer* appears attenuated as compared to the foregoing field. The outer line of Baillarger appears as a dense and broad stripe. The breadth of the external tenia is diminished, but clearly exceeds that of the internal tenia.

The *ganglionic layer* is broader than in the core, in particular the upper part, P_{Va}, is now a clearly recognizable stripe (Fig. 28). The inner tenia remains relatively narrow, showing a blurred upper and a sharp lower border. Besides its usual constituents, it contains a scarce population of Betz pyramids (see Chap. 7.4.1). The interstriate zone (5a) appears lightened. The inner line of Baillarger is of the same tint as the outer one.

The *multiform layer* remains unchanged.

Fig. 28. Area parietalis granulosa (*gran*) and paragranulosa (*paragran*). The sections ▶ have been successively cut from the same block of tissue. *Upper third* Nissl preparation (30 μm). *Middle third* Myelin preparation (100 μm). *Lower third* Pigment preparation (800 μm). Note the broad granular layer in the Nissl preparation, the conspicuous inner stripe of Baillarger (*Bi*) and the broad external tenia (*Te*) with blurred upper border as characteristics of the parvocellular parietal core (*gran*). The beginning of the paragranulous belt (*paragran*) is indicated by *triangles*

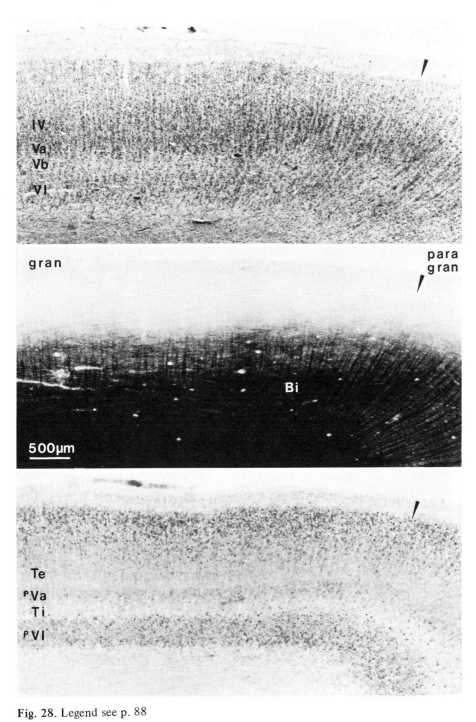

Fig. 28. Legend see p. 88

7.3.3 The Parietal Magnopyramidal Region

Posterior to the parietal paragranulous field there stretches out a vast territory of "association" areas. Close to the lower extremity of the belt there is the parietal magnopyramidal region which spreads over large parts of both the postcentral gyrus and the inferior parietal lobule (Fig. 34).

The broad parietal magnopyramidal cortex distinguishes itself from the belt by an enlarged inner main stratum. The corpuscular and pyramidal layers resemble those of the belt except for the great number of large pigment-laden IIIc-pyramids. The granular layer is split into an upper part devoid of pigment and a lower stripe richly stocked with faintly pigmented nerve cells (^{p}IVb). The outer tenia is narrow. Sublayer ^{p}Va appears broadened at the expense of the inner tenia which is in turn invaded by such a great number of pigmented pyramids as to almost defy recognition. The field displays therefore an equoteniate propebiteniate characteristic. The lines of Baillarger are hardly recognizable because of the dense distribution of fibres not only in the intrastriate zone but also in the substriate one. The region can nevertheless be classified with the equodense propebistriate type.

In sum, a narrow stripe of cortex along the anterior wall and the crest of the postcentral gyrus bears the somatosensory core and belt. The parietal "association" areas are particularly expanded. A large part of this territory is formed of the parietal magnopyramidal region.

The *parietal granulous core* strikes one by the smallness of its nerve cell constituents, its high myelin density (typus dives) and its pallor in pigment preparations (typus clarus). The cortex is extremely externocrassior and externoteniate. There is a predominance of the inner line of Baillarger.

The *parietal paragranulous belt* is far less granularized. It remains well endowed with myelinated fibres but shows an average pigment content. The field is externocrassior, modestly externoteniate, and equodense (bistriate).

The magnopyramidal region as part of the *"association" areas* is formed of homotypical isocortex. There is only average content of both myelinated fibres and pigment deposits. As to the Baillargers, the region is equodense and propebistriate. The teniae show the equoteniate and propebiteniate characteristic.

Lesions of the inferior parietal lobule which in all probability receives converging input from visual, auditory, and somatosensory cortical fields (Hyvärinen and Poranen, 1974; Mesulam et al., 1977) and a heterogenous thalamic input (Kasdon and Jacobson, 1978) can be followed by severe impairment of speech perception, spatial orientation, and sensory discrimination (Moffie, 1949; Hécaen et al., 1956; Geschwind, 1965; Petrides and Iversen, 1979).

7.4 The Frontal Lobe (Figs. 29, 30, 34)

Frontal Ganglionic Core: Campbell (1905, Fig. 35) posterior parts of the precentral area; Smith (1907, Fig. 36) posterior parts of area praecentralis A; Brodmann (1909, Fig. 37) posterior parts of area gigantopyramidalis (4); Vogt (1910, Fig. 38), Sanides (1962, 1963, 1964) area astriata typica (42); von Economo and Koskinas (1925, Figs. 39, 40) part of area gigantopyramidalis (FAγ); Bailey and von Bonin (1951, Fig. 41) isocortex agranularis gigantopyramidalis praecentralis; Sarkissov et al. (1955, Figs. 42, 43) part of field 4.

Frontal Paraganglionic Belt: Campbell (1905, Fig. 35) anterior parts of the precentral area and upper parts of the intermediate precentral area; Smith (1907, Fig. 36) parts of the area praecentralis A and area praecentralis B; Brodmann (1909, Fig. 37) anterior parts of area gigantopyramidalis (4); Vogt (1910, Fig. 38) area propeastriata subunistriata (38 and 39) and part of area unistriata degrediens dives (40); von Economo and Koskinas (1925, Figs. 39, 40) part of area praecentralis (FA) and area frontalis agranularis (FB); Bailey and von Bonin (1951, Fig. 41) part of isocortex agranularis simplex praecentralis; Sarkissov et al. (1955, Figs. 42, 43) part of field 6.

Inferofrontal Magnopyramidal Region: Campbell (1905, Fig. 35) lower parts of the intermediate precentral areas; Smith (1907, Fig. 36) area frontalis inferior B; Brodmann (1909, Fig. 37) posterior parts of area opercularis (44) and inferoanterior parts of area frontalis agranularis (6); Vogt (1910, Fig. 39) area propeunistriata (56) and part of area propebistriata (57); von Economo and Koskinas (1925, Figs. 39, 40) posterior parts of area frontalis intermedio agranularis magnocellularis in campo Broca (FBC_m); Bailey and von Bonin (1951, Fig. 41) part of isocortex dysgranularis frontalis; Sarkissov et al. (1955, Figs. 42, 43) part of field 44 and 6 op.

Superofrontal Magnopyramidal Region: Campbell (1905, Fig. 35) upper parts of the intermediate precentral and frontal area; Smith (1907, Fig. 36) area frontalis superior anterior; Brodmann (1909, Fig. 37) superior parts of area frontalis agranularis (6) and area frontalis intermedia (8); Vogt (1910, Fig. 38) area unistriata dives subtenuistriata et aequostriata (36 + 37); von Economo and Koskinas (1925, Figs. 39, 40) parts of area frontalis agranularis and area frontalis intermedia (FB and FC); Bailey and von Bonin (1951, Fig. 41) superior parts of isocortex agranularis simplex praecentralis; Sarkissov et al. (1955, Figs. 42, 43) superior parts of fields 6 and 8.

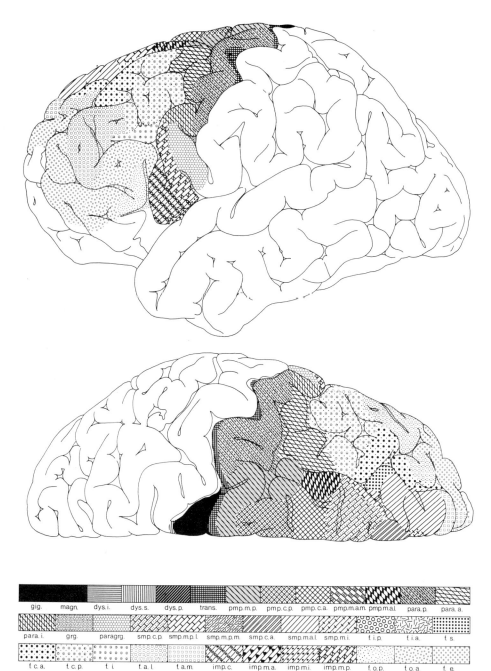

Fig. 29. Legend see p. 93

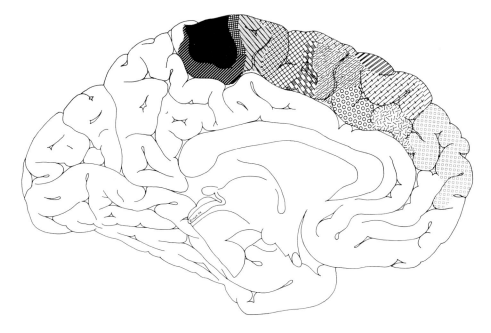

Figs. 29 and 30. Map of pigmentoarchitectonic areas of the human precentral, subcentral, and frontal regions. Note the location of area gigantoganglionaris (*gig*), area magnoganglionaris (*magn*), area dysganglionaris inferior, superior, and paracentralis (*dys.i, dys.s, dys.p*), area transganglionaris (*trans*), the inferofrontal magnopyramidal region (*imp.c., imp.m.a., imp.m.i., imp.m.p*), and the superofrontal magnopyramidal region (*pmp.m.p., pmp.c.p., pmp.c.a., pmp.m.al., pmp.m.am., smp.c.p., smp.m.pl., smp.m.pm., smp.c.a., smp.m.al., smp.m.i.*). Figure 34 does not show the individual areas but displays more clearly the location of the frontal magnopyramidal regions. For details concerning areas not described here see H. Braak (1979b)

There is now to add that also the frontal lobe exhibits a sequence of a refined core, a less specialized belt and a wide range of "association" areas, similar to that found in the occipital, temporal, and parietal lobe.

7.4.1 The Frontal Ganglionic Core

As opposed to the parvocellular sensory core fields, the somatomotor core is magnocellular in nature.

The posterior limit of the motor core lies along the anterior wall of the central sulcus close to its floor. On the medial facies of the hemisphere, the core is bounded by a line which is in continuation with the central sulcus coursing downwards to the upturned end of the cingulate sulcus whence it bends back and upwards at about the anterior limit of the paracentral lobule. Superolaterally, the borderline swings over to the free surface of a triangular part of the precentral gyrus whence it recedes into the central sulcus. The lower extremity of the core is reached at about the level of the inferior frontal sulcus (Figs. 29, 30).

In the subhuman primate brain the motor core occupies a by far more extended part of the free surface (Nañagas, 1923; Bucy, 1935; von Bonin, 1949; Rosabal, 1967).

The core receives a particular rich supply from the thalamic radiation. There is evidence for a precise somatotopical arrangement of motor foci not only in the motor core but also in the belt (Vogt and Vogt, 1926; Vogt, 1927; Penfield and Boldrey, 1937; Penfield and Rasmussen, 1950; Woolsey et al., 1952; Woolsey, 1955). Intracortical microstimulation reveals the existence of mosaic clusters of efferent nerve cells in the deep parts of the cortex. With phylogenetic advance these clusters become increasingly finely arranged (Asanuma and Ward, 1971; Asanuma and Rosén, 1972).

The hyperpyramidal motor core is signally broad. As one passes inferiorly the cortex narrows down continually. Also some other typical features disappear gradually and the cortex embodies more and more the traits of an unspecialized sensory-motor cortex. Amalgamation of both the somatosensory and the somatomotor features is to a variable extent generally found in the primitively organized mammalian brain (Lende, 1963; Lende and Sadler, 1967; Walsh and Ebner, 1970; Bohringer and Rowe, 1977).

The motor core of man can be subdivided into a posterior dysganglionic zone, an intermediate stretch of elaborated core fields (gigantoganglionic and magnoganglionic areas), and a transitory zone anteriorly (H. Braak, 1979b). It seems advisable to choose the gigantoganglionic area for an introduction of the distinguishing features of the motor core.

The Gigantoganglionic Area (Figs. 27, 31). The cell-sparse *molecular layer* contains a fair number of myelinated fibres located immediately subjacent to the external glial zone.

The *corpuscular layer* is dominated by tightly packed small pyramids and a scattered population of pigment-laden stellate cells.

The *pyramidal layer* accommodates pyramids which form a gradient with the largest ones at the lower border. The amount of pigment stored

---

Fig. 31. Area gigantoganglionaris of man. *Upper half* Nissl preparation (15 μm). *Lower half* Myelin preparation (15 μm) *left,* and pigment preparation (800 μm) *right.* Note the magnocellular, agranular, and internocrassior character of the field in the Nissl preparation, and the markedly externoteniate character in the pigment preparation. The population of unusually large and pigment-laden pyramids in layer Vb catches the eye

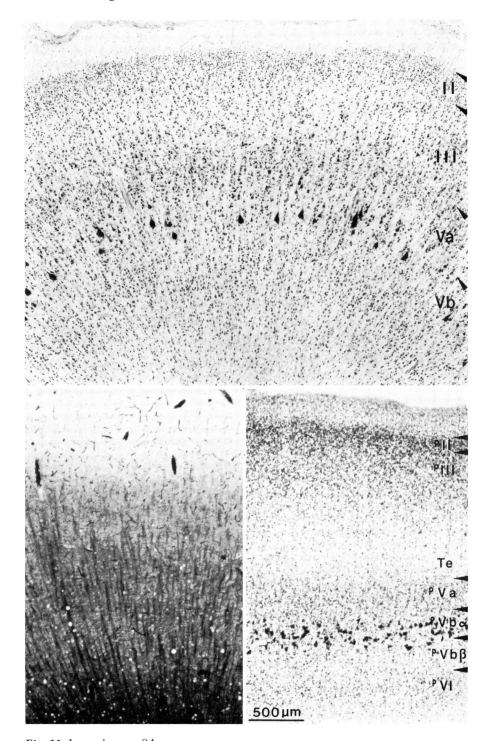

Fig. 31. Legend see p. 94

in the individual pyramids is subjected to a gradual decrease from above downwards. Hence, the large pyramids are almost devoid of lipofuscin deposits, a pattern which invariably persists as age advances. The external tenia covers the lower reaches of the pyramidal layer. It is extremely broad and displays a strikingly blurred upper border. Also the lower one is by far not as sharply traced as in the postcentral core. The position of the outer tenia corresponds with that of the outer stripe of Baillarger.

A *granular layer* cannot be outlined in the motor cortex of the adult. It is apparently present during foetal life (Brodmann, 1903, 1905).

The *ganglionic layer* can clearly be subdivided into a cell-rich and modestly pigmented Va and a cell-sparse Vb, each of approximately the same breadth. Sublayer Va contains in abundance small-sized pyramids which tend to invade both teniae. Occasionally, it is referred to as the "internal granular" layer (Gatter and Powell, 1978).

Pigment preparations permit subdivision of sublayer Vb into an upper part $^PVb\alpha$ which accommodates the conspicuous Betz cells and a cell-sparse $^PVb\beta$ (Figs. 27, 31). In sharp contrast to the common type of pigment distribution in pyramidal cells the Betz cells amass lipofuscin granules in blocks (see Chap. 3.3). Due to the presence of large Betz cells the cortex gains an extremely internocrassior character.

Betz cells differentiate early. As a result of the degree of development of their cellular processes, they appear the most advanced cells of the motor cortex in the new-born (Conel, 1939–67).

The homogeneously pigmented *multiform layer* is broad and merges with the white substance.

Both the ganglionic and the multiform layer are filled with a dense plexus of myelinated fibres. The lines of Baillarger therefore normally defy recognition; the area displays an astriate character (Vogt, 1910). The outer line fills up sublayer IIIc, the inner one Vb, as can be visualized in suitably differentiated preparations (Vogt and Vogt, 1919).

The Magnoganglionic Area. The magnoganglionic core field is buried in the depth of the central sulcus. It differs from the gigantoganglionic area in that both the average size and the packing density of Betz cells are considerably reduced. Also the overall density of pigmentation is markedly decreased.

The Inferior Dysganglionic Area. The inferior dysganglionic field displays an extreme stage in the diminution and rarefication of the Betz cells. $^PVb\alpha$ appears as an almost empty layer where Betz cells can only occasionally and irregularly be encountered.

The Superior Dysganglionic Area. A relatively high packing density of markedly small-sized Betz cells is the main feature of the superior dysganglionic field (Fig. 27).

The Paracentral Dysganglionic Area. The paracentral dysganglionic area covers a wide territory of the paracentral lobule. $^PVb\alpha$ is filled with Betz cells of particularly small size. The average pigment content of the cortex is well above that found in the other dysganglionic areas.

The Transganglionic Area. The posterior border of the transganglionic area is given by the appearance of a densely pigmented sublayer ^PVIa which is absent in the remaining parts of the motor core. The anterior border is defined by the disappearance of the Betz cells.

7.4.2 The Frontal Paraganglionic Belt

Starting out on the medial facies of the hemisphere, the borderline of the paraganglionic belt can be followed along the upper wall of the cingulate sulcus whence it curves upwards at about the level of the anterior commissure. Superolaterally, it descends in an oblique course, reaching the precentral gyrus at the height of the inferior frontal sulcus where it abuts on the subcentral region (Figs. 29, 30).

Parallel with the core, the belt attains its largest spread dorsomedially and narrows rapidly as its inferior extremity is approached. From the upper margin downwards the features of the belt become gradually indistinct.

As compared to the core, the breadth of the cortex is slightly reduced in the belt which is furthermore devoid of Betz cells. On closer examination it becomes apparent that by and large the pyramids in layers III and V are smaller than those in the core. The internocrassior characteristic is less expressed. In Nissl preparations, the lamination is better recognizable. Pigment preparations show only a weak preponderance of the external tenia. The pyramidal cells of ^PIII, ^PVa, and ^PVI are on the average more densely pigmented than those in the core fields. The multiform layer is split into a band-like ^PVIa and a broad and inconspicuous ^PVIb which is continuous with the homogeneous ^PVI of the core. The bipartition of the multiform layer is a hallmark of the motor belt and areas in front of it. To all appearances, the heavily pigmented nerve cells of ^PVIa cannot be found in other parts of the telencephalic cortex.

In sum, the general features of the ganglionic *core fields* appear to be the large average size of the nerve cells, their low packing density, the preponderance of pyramids, the particularly rich supply with myelinated fibres (typus dives) and the pallor in pigment preparations (typus clarus). The astriate fields are extremely internocrassior and externoteniate.

The *belt areas* are a trifle less magnocellular in character and their internocrassior characteristic seems weakened. There is a slight reduction in overall myelin density; especially the tint of the substrate zone is heightened, leading to a propeastriate characteristic. The belt areas are only modestly externoteniate and show a clear increase in pigmentation.

The *"association" areas* in front of the core and the belt generally show an equoteniate characteristic. The granular layer becomes more and more clearly recognizable.

7.4.3 The Frontal Magnopyramidal Regions

Anterior to the paraganglionic belt there is an expanded territory of "association" areas. The following description concentrates on magnopyramidal areas which can be encountered within this stretch of cortex. Additionally, some of the belt areas themselves are magnopyramidal in character. Two magnopyramidal regions can be distinguished in the frontal lobe.

7.4.3.1 The Inferofrontal Magnopyramidal Region

The inferofrontal magnopyramidal region is relatively small. It lies immediately in front of the lower extremity of the belt spreading over basal parts of the fronto-parietal operculum. The anterior border of the region generally coincides with the course of the diagonal sulcus (Smith, 1907; Connolly, 1950).

The appearance of large and slender pigment-laden IIIc-pyramids is the special feature which serves to distinguish the region from all the areas surrounding it. As in the temporal lobe, the region separates into a central field with abundant pigment-laden IIIc-pyramids and fringe areas with a rarefied population of this type of pyramidal cell (H. Braak, 1979b).

The Central Magnopyramidal Field. Subjacent the cell-sparse *molecular layer* there is a *corpuscular layer* formed of a wealth of small pyramids and stellate cells.

The *pyramidal layer* is dominated by large pyramids, the size of which exceeds that of the fifth-layer pyramids (Fig. 33, arrow-head). In contrast to the precentral motor fields, the layer displays a clear-cut lower border in pigment preparations.

The conspicuous pigment-laden IIIc-pyramids are irregularly dotted about both the lower reaches of P_{III} and the external tenia.

The *granular layer* is ill-defined in Nissl preparations.

The *ganglionic layer* is dominated by large pyramids. It is split into a markedly well-pigmented ^{P}Va and a pallid ^{P}Vb. ^{P}Va is more deeply tinged than ^{P}III. Due to its clear-cut borderlines it attains a band-like appearance. Both teniae are equally wide. Myelin preparations display a propeunistriate internodensior characteristic.

The broad *multiform layer* fades into the white substance. Pyramids prevail in this field. The cortex is externocrassior, internodensior, and equoteniate in character.

The Speech Centre of Broca. The territory covered by the inferofrontal magnopyramidal region has long been considered to represent the motor speech centre of Broca (Brodmann, 1909, 1912, 1914; von Economo and Koskinas, 1925; Rose, 1935). Attempts to delineate the centre of Broca (1861, 1863) based on stimulation experiments and postmortal examination of brains from aphasic patients (Schiller, 1947; Penfield and Rasmussen, 1950; Russell and Espir, 1961; Benson, 1967; Benson and Patten, 1967; Hécaen and Consoli, 1973; Rasmussen and Milner, 1975; Kertesz et al., 1977; Ojemann and Whitaker, 1978) are in concert with the location and expansion of the inferofrontal magnopyramidal region.

7.4.3.2 The Superofrontal Magnopyramidal Region

The superofrontal magnopyramidal region is rather extended and covers considerable parts of the first frontal convolution (H. Braak, 1979b). It is in the posterior part composed of paraganglionic belt areas (Figs. 29, 30).

The region separates into central fields distinguished by an abundance of pigment-laden IIIc-pyramids and fringe areas with less densely packed nerve cells of this type.

Paraganglionic Belt: The Central Magnopyramidal Field. The molecular and corpuscular layer remain much the same as in the paraganglionic belt.

The *pyramidal layer* is richly stocked with slender and well-pigmented pyramids. The outer tenia is hardly recognizable because of the rich store of pigment-laden IIIc-pyramids. Its breadth clearly exceeds that of the inner one.

The *ganglionic layer* is split into a feebly tinged upper half with blurred boundaries and a lower one with a loose distribution of sparsely pigmented nerve cells. The outer tenia and ^{P}Va are almost equally wide.

The bipartited *multiform layer* shows a conspicuous band-like ^{P}VIa rich in pigment.

Superofrontal Region: The Central Magnopyramidal Field. As one goes polewards, the cortex as a whole narrows down. The breadth of the outer cellular laminae (II + III) increases at the expense of the deeper ones.

There is a marked increase in pigmentation of the pyramidal layer which more and more displays a sharply drawn lower border in pigment preparations. The breadth of the narrow outer tenia does not exceed that of the inner one. The upper part of the ganglionic layer (^{p}Va) is almost twice as broad as the outer tenia. The bipartition of the multiform layer is poorly expressed.

Magnopyramidal Regions and Regional Cortical Blood Flow. In all probability, the various magnopyramidal regions of the human brain are well displayed in the studies on regional cortical blood flow (Fig. 32). As a rule, blood flow depends on the rate of oxidative metabolism in the non-anoxic brain. Changes in the activity of the nerve cells are therefore reflected by corresponding changes in the blood flow (Ingvar, 1975, 1978; Raichle et al., 1976). Recently developed techniques allow for the precise demonstration of the regional cortical blood flow at rest or during the performance of different tasks (Ingvar and Schwartz, 1974; Ingvar, 1976; Roland and Larsen, 1976; Ingvar and Philipson, 1977; Larsen et al., 1978; Lassen et al., 1978a,b).

The cortex in front of the central sulcus shows generally a higher rate of blood flow than that behind this landmark. In particular, the upper part of the superior frontal convolution is well supplied (Risberg and Ingvar, 1973; Ingvar, 1976; Larsen et al., 1978; Lassen et al., 1978b; Roland et al., 1980). Location and extension of this zone activated already during rest corresponds closely to the superofrontal magnopyramidal region (Fig. 32).

During the execution of simple automatic speech tests this particular zone shows a marked increase in blood flow (Ingvar, 1978; Larsen et al., 1978). Also clinical data suggest that in addition to the inferofrontal magnopyramidal region parts of the superior frontal convolution which lie well within the limits of the superofrontal magnopyramidal region are involved in the performance of speech (Penfield and Rasmussen, 1949, 1950; Erickson and Woolsey, 1951; Penfield and Roberts, 1959; Talairach and Bancaud, 1966). Parasagitally localized tumours, disturbances of the anterior cerebral artery circulation, or surgical excision of parts of the first frontal convolution may cause aphasia even if the lower frontal convolution on both sides is well preserved (Chusid et al., 1954; Petit-Dutaillis et al., 1954; Guidetti, 1957; Arseni and Botez, 1961; Carrieri, 1963; Rubens, 1975; Laplane et al., 1977; Masdeu et al., 1978). Also extended injuries or even bilateral extirpation of Broca's area may only result in a mild and transient form of aphasia (Mettler, 1949; Zangwill, 1975). All these experiences buttress the presumption that the frontal lobe harbours *two* centres for the motor control of muscles involved in speech performance.

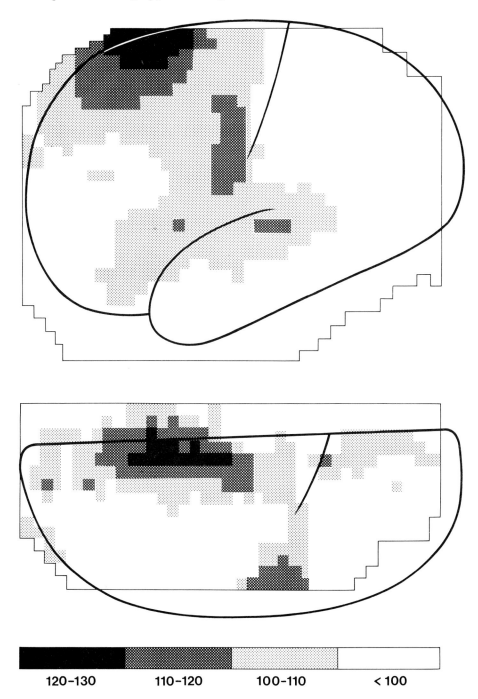

Fig. 32. Location of cortical regions showing a marked increase in blood flow during the execution of simple automatic speech tests. The superofrontal magnopyramidal region shows a particularly high blood flow, but also the inferofrontal magnopyramidal region and the temporal magnopyramidal region appear activated. Simplified and redrawn from Larsen et al. (1978)

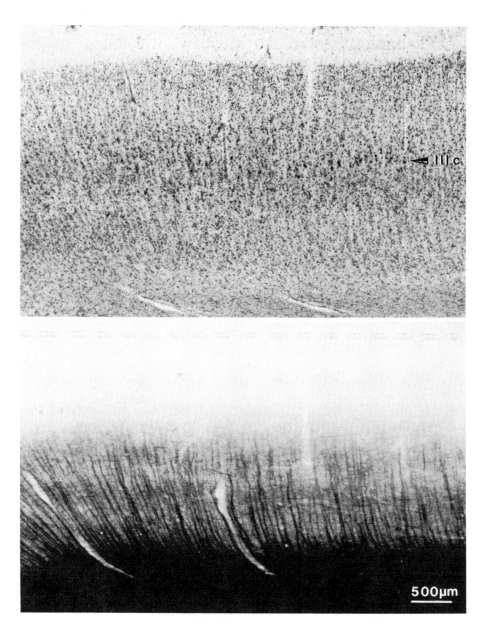

Fig. 33. Area inferofrontalis magnopyramidalis centralis of man. The Nissl and myelin preparation (15 μm) have been successively cut from the same block of tissue. Note the large IIIc-pyramids (*arrow-head*) and the thick radii

Fig. 34. Map of the major regions of the human isocortex. *ag* anterogenual core; *cm* cingulate magnoganglionic core; *e* entorhinal region; *fb* frontal paraganglionic belt; *fgc* frontal ganglionic core; *ifm* inferofrontal magnopyramidal region; *ob* occipital paragranulous belt; *ogc* occipital granulous core; *om* occipital magnopyramidal region; *pb* parietal paragranulous belt; *pg* paragenual belt; *pgc* parietal granulous core; *pl* paralimbic areas; *pm* parietal magnopyramidal region; *ps* parasplenial belt; *rs* retrosplenial core; *s* subcentral region; *sfm* superofrontal magnopyramidal region; *tb* temporal paragranulous belt; *tgc* temporal granulous core. The regions indicated by fine stippling are not dealt with in detail

The Superofrontal Magnopyramidal Region

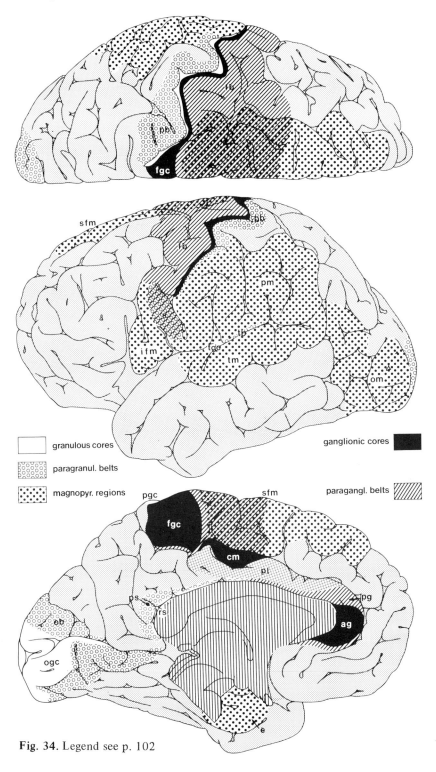

Fig. 34. Legend see p. 102

8 Brain Maps

A considerable number of cortical maps of the human brain have been drawn up by different authors. For the most part they yielded to the temptation to summarize their results in a brief way. But even the most elaborately constructed map gives only a rough idea of the architectural plan, since it shows only the free surface and does not display the almost two-thirds of the cortex which is hidden away in the depths of the sulci. It is very difficult to get even a rough idea of the extent and boundaries of cortical areas from these brain maps. This shortcoming becomes particularly apparent if one attempts to compare cortical mappings of different authors with one's own results.

Fig. 35 displays the first cortical map of the human brain which was developed by the Australian Alfred Walter Campbell (1868–1937). His investigations were based on serial sections stained for nerve cells and myelin sheaths. Campbell arrived at a subdivision of the cortex into 14 areas (1905). On closer examination it becomes apparent that the map displays areas which are far from being homogeneous in structure. His areas should therefore be more aptly considered cortical regions, the subdivisions of which may partly be structurally related to each other.

The allocortex is poorly described in Campbell's map. It is displayed as a uniform "olfactory" area despite the fact that its major parts are strikingly different from each other in both the Nissl and the myelin preparations. There is also no distinction made between the anterior and posterior territories of the cingulate gyrus which again are fundamentally different from each other (magnocellular and poor in myelin versus parvocellular and rich in myelin).

As concerns the isocortex, Campbell does not recognize the structural differentiations within the occipital cortex outside the striate area. The scheme of the superolateral facies of the brain gives only a poor representation of the temporal lobe. Campbell's "temporal" area spreads not only over the middle and inferior temporal convolutions but covers also major parts of the inferior parietal lobule. It is evident that Campbell did not pay attention to the appearance or disappearance of unusually large IIIc-pyramids which clearly set off the cortex of the inferior parietal lobule from that of the middle and inferior temporal convolutions. The border between the precentral and the intermediate precentral field corresponds to that of area 4 and 6 of Brodmann. The surface spread of the precentral field is exaggerated. The intermediate precentral field encloses Brodmann's areas 6, 44, and 45, although Nissl preparations clearly reveal the agranular nature of area 6 as opposed to the granular character of fields 44 and 45

Brain Maps

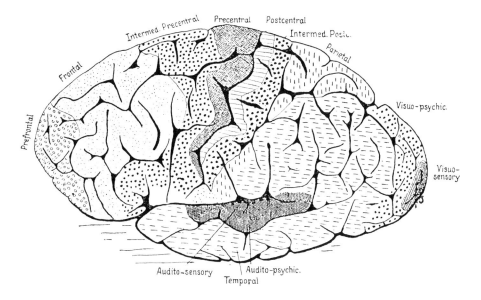

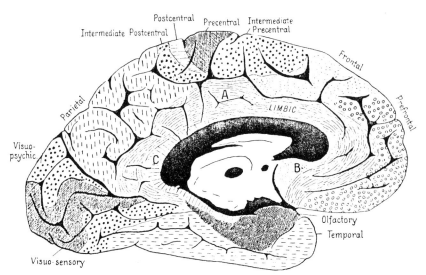

Fig. 35. Legend see p. 104

Fig. 36. The beginnings of myeloarchitectural analysis of the telencephalic cortex can ▶
be traced back to Gennari (1782), Vicq d'Azyr (1786), and Baillarger (1840), who
examined the whitish stripes of myelinated intracortical fibres in unstained preparations.

Such attempts to study the cortex macroscopically reached their peak in the work
of the Australian Grafton Elliot Smith (1871–1937). He made use of the local variations of the two stripes of Baillarger, which can be recognized even with the unaided
eye. Smith finally arrived at a subdivision of the human cortex into about 50 areas.
His map, which appeared in a short paper in 1907, is in many respects superior to that
of Campbell. Although Smith was quite aware that he was apt to overlook real differences by limiting his investigations to only the gross appearance of cortical myelinated
fibres, his work clearly illustrates the value of myeloarchitectonic studies. As a rule,
even the unexperienced investigator is capable of delineating cortical areas with the
aid of myelin preparations. Moreover, macroscopical or low-power microscopical
examination is actually an appropriate way of studying the architectonics of the brain
since it permits easy recognition of principal cortical variations. The resolution is not
so great, however, that one necessarily sees the subtle and confusing differences apparent in Nissl preparations. Analysis is therefore much easier.

In the map of Smith the allocortex is much less pertinently parcellated than the
isocortex. There is, for instance, no substantiation to unite as he did the basal parts of
the frontal lobe with adjacent ones of the insula and the entorhinal region.

The main territories of the isocortex are clearly delineated. Smith has the merit of
distinguishing a peristriate territory from a parastriate area which immediately surrounds the visual core. The extent of the acoustic core appears exaggerated, an error
which is understandable since the areas which accompany the acoustic core are almost
as heavily myelinated as the core itself. The cortex covering the superior temporal gyrus
is sharply set off from that spreading over the subjacent gyri. The latter is described
as an almost uniform territory showing only variations in the total thickness of the
cortex. Smith points already to the fact that isocortical areas are for the most part
bounded by relatively sharp borderlines

Brain Maps

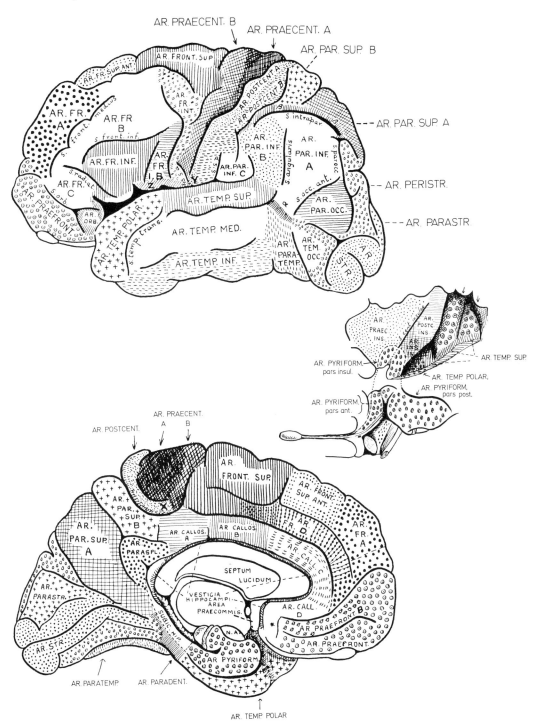

Fig. 36. Legend see p. 106

Fig. 37. The most famous cortical map of the human brain is that of Korbinian Brodmann (1868–1918) who beginning in 1901 worked together with the Vogts in Berlin. Brodmann based his parcellation on the analysis of Nissl preparations, whereas the Vogts mainly concentrated on myeloarchitectonic studies. Brodmann extended his studies to include a great number of subhuman primate brains and even subprimate mammalian brains. After eight years of intensive morphological investigations, he summarized his experience in a relatively short exposition on the human brain (1908) and a more comprehensive monograph (1909), both containing the world-famous map.

It is deemed unfortunate, however, that Brodmann's accounts on the human telencephalic cortex (1908, 1909, 1912, 1914) do not contain descriptions of the structural peculiarities of the fields outlined and give only short comments concerning their topographic localization. The studies also lack photographic illustrations, so that difficulties arise if one attempts to compare Brodmann's map with those of other authors. A comprehensive atlas on the human brain had been in progress in the institute of the Vogts but unfortunately the work on it was interrupted after the appearance of the atlas of von Economo and Koskinas (1925).

In the human brain, Brodmann distinguished 44 cortical areas and classed them in 11 regions. In general, he numbered the areas in the order in which they appeared when investigating serial sections of smaller blocks from the brain cut in suitable planes, i.e., almost horizontally in cases of fields 1–7 but transversally in those of fields 17–19. There are gaps in the counting sequence between the numbers 12 to 16 and 48 to 50. The frontopolar area 12 is found only in the subhuman mammalian brain. As is depicted in Fig. 37, the insular region is divided into an agranular anterior portion and a granular posterior one. In the brains of subhuman primates, Brodmann designates the granular part as area 13 from which he sets off the agranular areas 14 to 16. The areas 48 to 50 can only occasionally be encountered in subhuman mammalian brains. Area 51 is not labeled in Fig. 37. It summarizes the rudimentary cortex of the prepiriform region and the olfactory tubercle.

The small sketch showing the insular and the supratemporal plane was published in 1909, whereas the figures of the superolateral and the medial facies of the brain represent Brodmann's latest map, which appeared in 1914. It differs slightly from the earlier map in that it shows subdivisions of area 7 into 7a and 7b, and of area 44 into 44 and 44o. Additionally, the area 52 has been inserted.

The Vogts arrived at a far greater number of cortical areas in their myeloarchitectonic studies. They nevertheless used for sake of perspicuity Brodmann's system of numbering of cortical areas when they hypothetically transferred results of experimental stimulation of the monkey's brain on the surface relief of the human brain (1919, 1926). Unfortunately the resulting map displays severe misinterpretations such as the designation of the inferior parietal lobule as area 7. This mistake can also be encountered in Foerster's map in which his results of stimulation experiments of the human brain executed during neurosurgery have been summarized (1936).

Major parts of the allocortex such as the fascia dentata, the cornu ammonis, and the subiculum do not appear in Brodmann's system. Only the presubiculum (area 27) and the entorhinal region (areas 28 and 34) are delineated. The complex transition from the entorhinal allocortex to the temporal isocortex is not recognized but there is an ill-defined perirhinal field (area 35) inserted. Parcellation of the cingulate gyrus is by far better than in the foregoing maps. It is one of the merits of Brodmann that he calls particular attention to the intricate construction of the retrosplenial region.

The correct distinction between the visual belt area (area 18) and the adjoining cortex (area 19) is another highlight in the history of cortical parcellation, since the border between areas 18 and 19 is relatively easily overlooked in Nissl preparations. The acoustic core and belt are by and large correctly delineated. The territory of transition between the occipital, the parietal, and the temporal lobes is still a matter of debate (area 37). Brodmann considers area 37 as belonging to the temporal lobe. There is no clear description substantiating Brodmann's distinction between areas 20 and 21. The surface spread of both the postcentral area 1 (the paragranulous belt) and the precentral area 4 (the ganglionic core) remains exaggerated. Brodmann also did not realize that the mature parietal isocortex extends into the free surface of the cingulate gyrus

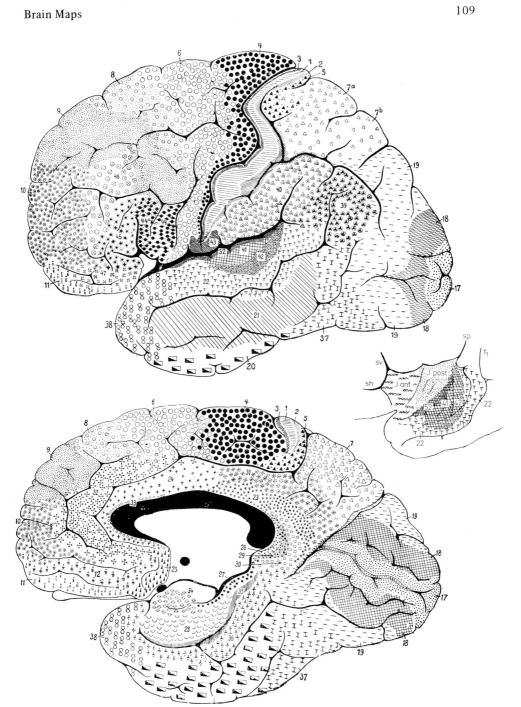

Fig. 37. Legend see p. 108

Fig. 38. The development of staining procedures for myelin sheaths stimulated greatly ▶ the further architectural analysis of the telencephalic cortex. Unfortunately, only a few techniques can be used for a sharp and almost selective staining of intracortical fibres (see Chap. 8). These methods require frozen sections which are difficult to handle if cutting of large blocks seriatim is attempted. A further shortcoming is the necessity of a differentiation step which remarkably influences the final staining result.

This figure displays a composite drawing of the maps of the Vogts which have been published in different papers (C. and O. Vogt, 1919; O. Vogt, 1910, 1911, 1927; the composite drawing is taken from the textbook of von Economo and Koskinas, 1925). The Vogts did not arrive at a parcellation of the temporal lobe and the occipital lobe. Their map is the only incomplete one reprinted here. One of the main characteristics of the map is the strikingly large number of cortical areas outlined. Brodmann's fields reappear as extended territories composed of numerous architectural entities. Already about a hundred cortical areas are delineated in the Vogts' subdivisions of the frontal and the parietal lobe. The existence of these areas has found corroboration in numerous studies by subsequent observers using either Nissl or myelin preparations or both (Strasburger, 1937; Gerhardt, 1940; Batsch, 1956; Sanides, 1962). Pigment preparations alone do not provide a basis for such an extensive parcellation as proposed by the Vogts. This might partly be due to shortcomings of the method, since the staining technique is deliberately used with unusually thick sections which not only efface subtle local differences but also bring into prominence the constant variations of the cortex.

We owe to the Vogts an accurate description of most of the allocortical areas. Also the proisocortical regions of the cingulate gyrus have repeatedly found their interest as zones of transition from the allocortex to the isocortex. As concerns the free surface, the maps of the Vogts show for the first time a realistic outline of the precentral and the postcentral core fields

Brain Maps

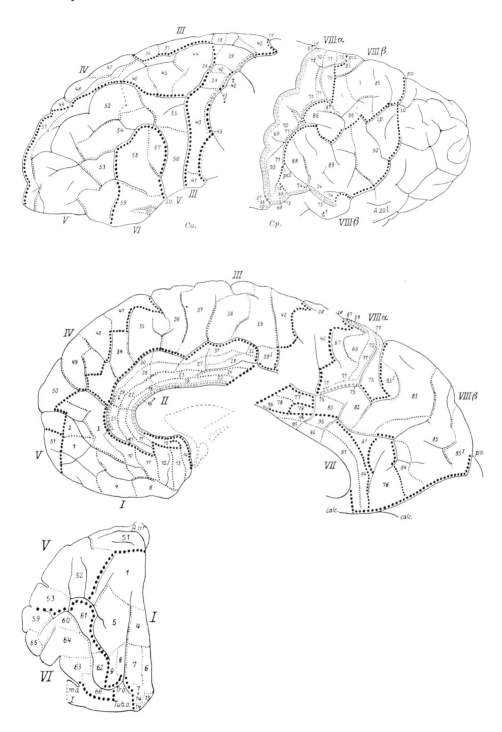

Fig. 38. Legend see p. 110

Figs. 39 and 40a,b. In 1925, Constantin von Economo (1867–1931) published in cooperation with Koskinas a comprehensive textbook and atlas on the architectonics of the telencephalic cortex of the human adult. The book can be considered the result of 12 years of intensive morphological investigations and contains not only painstaking descriptions of each of the various fields delineated but also illustrative photomicrographs of all these areas. It thereby compensated for the shortcomings of Brodmann's analysis. Von Economo arrived at a parcellation which is by and large comparable to that given by Brodmann. 57 cortical areas are distinguished but additional descriptions of 50 modifications are given which are not outlined in the map but obviously can be considered as real architectural entities.

The drawings of the brain are to a certain extent simplified and display idealized courses of both the gyri and sulci. The opercula of the lateral cerebral sulcus are pushed upwards to disclose the areas covering the insula and the supratemporal plane.

With the exception of isocortical core fields, von Economo deliberately avoids drawing sharp borderlines between individual areas. Moreover, the author emphasizes the existence of more or less broad zones of transition between individual areas. This has led to severe controversy with the Vogts who, by contrast, claimed hairsharp borderlines between cortical areas (Vogt, 1928). In this book the reader has become acquainted with not only sharp borderlines (border of the striate area) and relatively clear-cut ones (border of the acoustic core field) but also with hazy boundaries which show within narrow limits a gradual transition from one cortical field to another (anterior border of the transganglionic field, peripheral border of most of the magnopyramidal areas) and rather complex territories of transition with interdigitating laminae as in the transentorhinal and transsubicular regions. Hence, there is no reason to claim the existence of hairsharp borders for all cortical fields. The constructions of the individual borders appear rather as manifold as the constructions of the areas themselves.

When studying cortical architecture, von Economo used to dissect the hemispheres into a great number of blocks in order to cut the cortex as far as possible perpendicular to the surface. This processing is linked with considerable loss of tissue and has therefore been heavily criticized by the Vogts. They emphasized the superiority of uninterrupted serial sections cut in either the transversal, the horizontal, or the sagittal plane.

In the map of von Economo particular attention is paid to the retrosplenial proisocortex. Posterior parts of the cingulate gyrus (areas LC and LD) which display features of the mature parietal isocortex are classed with the "limbic lobe", an assailable labelling which probably reflects the influence of Brodmann's map. The territory of transition between the parietal, the temporal, and the occipital lobe (area PH) is considered as being formed of parietal cortex. This interpretation has not found corroborative support in follow-up examinations. The cortex covering the middle and inferior temporal gyri is shown as an almost uniform field (TE) but again modifications TE1 and TE2 corresponding to Brodmann's areas 21 and 20 are distinguished. As concerns the free surface spread of the pre- and postcentral core and belt areas, the map repeats the errors of that of Brodmann.

Brain Maps

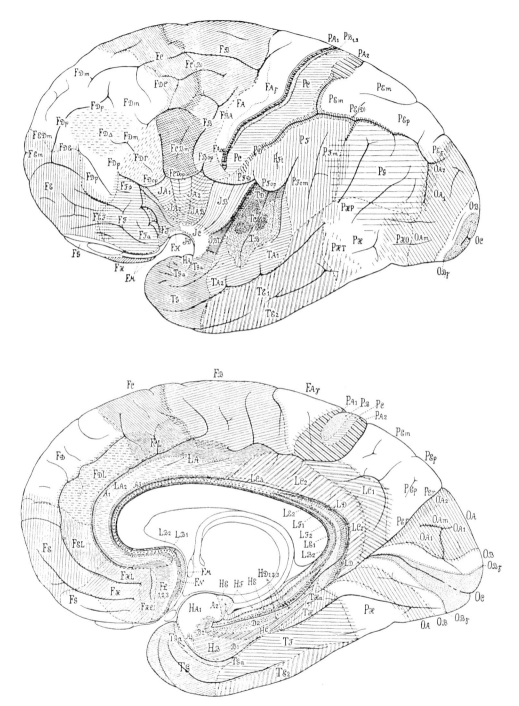

Fig. 39. Legend see p. 112

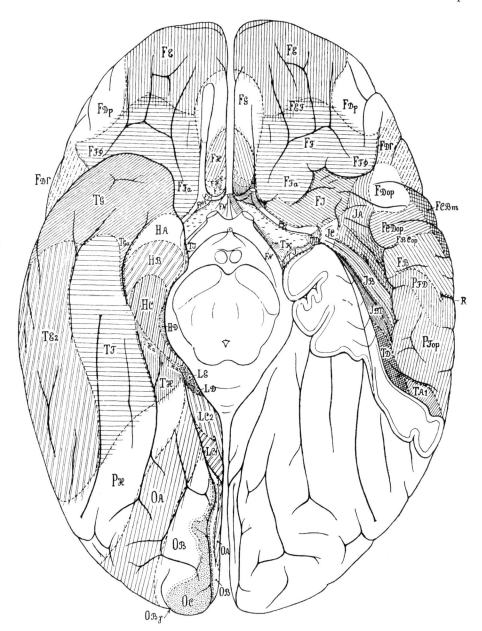

Fig. 40a. Legend see p. 112

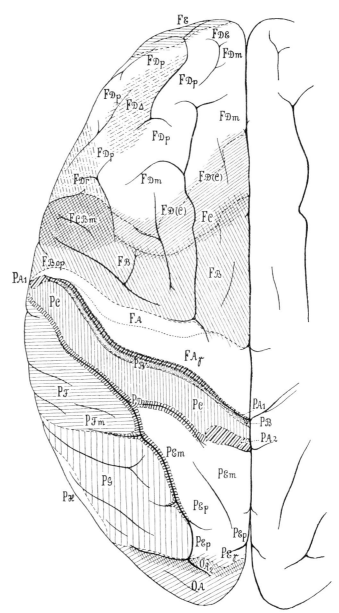

Fig. 40b. Legend see p. 112

Fig. 41. It was left to Percival Bailey and Gerhardt von Bonin (1951) to summarize their experience in cortical architectonics in a map disclosing a uniform "eulaminate" cortex that stretches over expanded territories of not only the frontal but also the temporal, parietal, and occipital lobes. The map, which originally appeared in colour, is here reproduced in the black and white version designed by Chusid (1964). Only parts of the allocortex and the isocortical core fields are shown with relatively sharp borders. Bailey and von Bonin accepted to some extent the existence of belt areas. But they claimed that these zones of transition merge continually with the adjoining "eulaminate" cortex. Although it may be in fact difficult to outline architectural entities outside the core and belt areas in Nissl preparations, there is obviously no reason to claim the existence of such an expanded stretch of homogeneously formed cortex.

The differences between the cortex covering the superior temporal gyrus and that spreading over the subjacent gyri may serve as an example (see Chap. 7.2.5). Betz had already in 1881 emphasized the existence of such differences. His observations have been accepted and corroborated in several follow-up examinations grounded on cyto-, myelo-, and pigmentoarchitectonics and need therefore no revision (Campbell, 1905; Brodmann, 1908; von Economo and Koskinas, 1925; Hopf, 1954, 1955, 1968; H. Braak, 1978b). The reason why Bailey and von Bonin deny such differences is rather enigmatic and the only plausible explanation for their proposal of the "eulaminate" cortex is that the authors contended themselves with superficial observations. To give an example: Plate I in the book is designated "Allocortex praepiriformis"; the figure nevertheless shows clearly a part of the entorhinal region. This is hard to explain, since even the unexperienced investigator is generally capable of making a distinction between the prepiriform and the entorhinal region (Stephan, 1975)

Correlation of colours used by Bailey and von Bonin (1951) with black-and-white equivalents (Chusid, 1964)

Colour	Area	Black-and-white
Pink	Homotypical isocortex	Small dots
Brown (pink and black)	Parakoniocortex	Small dots, shaded background
Red	Koniocortex	Small dots, dense
Yellow	Agranular cortex	Large dots
Gray (yellow and black)	Agranular gigantopyramidal cortex	Large dots, shaded background
Orange (yellow and pink)	Dysgranular cortex	Large and small dots
Blue	Allocortex	Vertical bars
Purple (pink and blue)	Juxtallocortex	Vertical bars and small dots
Green (blue and yellow)	Mesocortex	Vertical bars and large dots

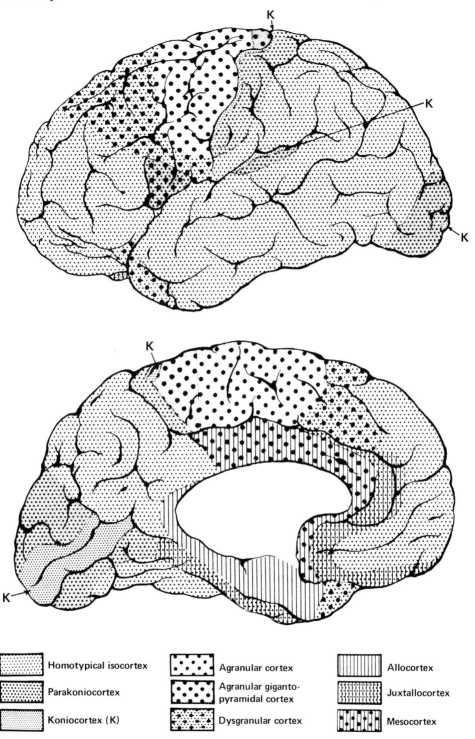

Fig. 41. Legend see p. 116

Figs. 42 and 43. The last complete map of the human telencephalic cortex is that of Sarkissov and co-workers, which appeared in 1955. The map can be considered to be merely an improved revision of that of Brodmann. The corrections can partly be traced back to the results of the myeloarchitectural studies of the Vogts; the organization of the anterior cingulate region and the restricted surface spread of area 4 may serve to give just two examples. It might be of interest in this connection to know that Oskar Vogt has been entrusted with the creation of the state institute of brain research in Moscow in 1925. During the time he spent in Moscow Vogt trained a group of young Russians in the field of architectonics. In the following period the institute grew up to an important centre of brain research. Both Sarkissov and Filimonoff, co-authors of the brain map under consideration, had formerly been pupils of Oskar Vogt, a fact which goes a long way toward explaining why the present map appears to some extent as a synopsis of Brodmann's and Vogt's work.

The textbook contains short descriptions in Russian of each of the fields outlined. An English translation is not available. The accompanying atlas is remarkably useful and in contrast to that of von Economo and Koskinas relatively easily accessible. It contains more than 200 conclusive and brillant illustrations of the various cortical areas outlined in the map. Most of them are reproduced at a magnification of 100:1. The location of the piece of cortex shown in the various figures is indicated by a rectangular frame in a small inset sketch of the complete section of the hemisphere. This is a great help in the laboratory when comparing the plates with own serial sections.

The work of Sarkissov et al. appears in summary as the coronation of the architectural work initiated by Brodmann and the Vogts, who as mentioned already unfortunately did not arrive at such a complete description and illustration of the human telencephalic cortex. The atlas may therefore well serve as a complement to the elaborate studies of the Vogts and their co-workers

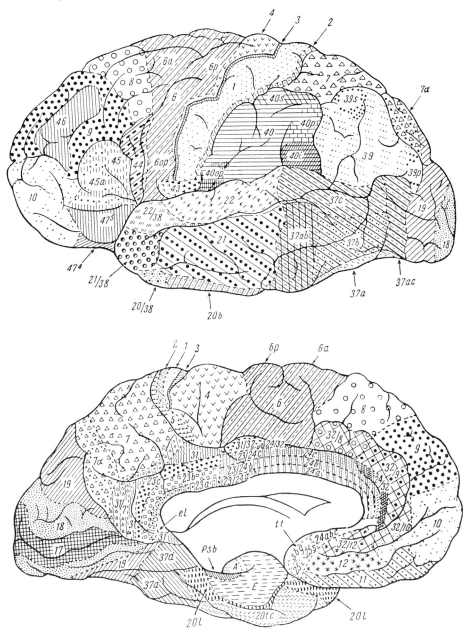

Fig. 42. Legend see p. 118

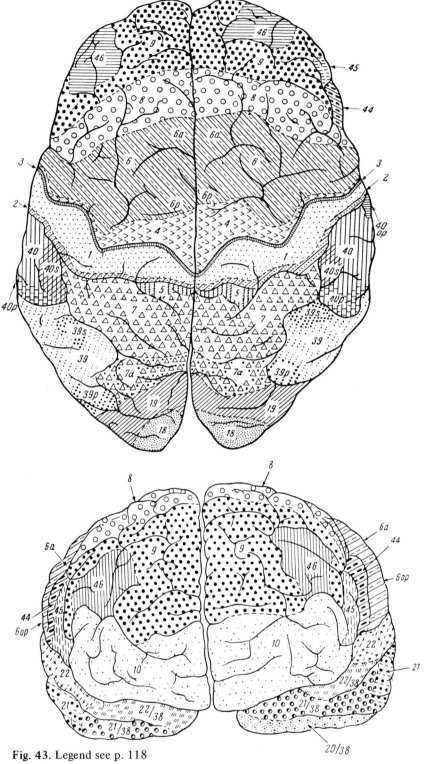

Fig. 43. Legend see p. 118

9 Notes on Techniques

For *cytoarchitectural* studies paraffin sections of formalin-fixed material, cut at 15 µm, and stained after the following formula are recommended.

1. De-wax and transfer through a graded series of ethanol to aqua dest.
2. Stain in a solution of cresyl violet for 30 min at 50°C.
   Stock solution A: 0.2% aqueous solution of cresyl violet (Chroma).
   Stock solution B: 2.72% aqueous solution of sodium acetate.
   Stock solution C: 1.20% aqueous solution of acetic acid.
   Shortly before use prepare staining solution by adding 50 ml solution A to 190 ml solution B plus 760 ml solution C.
3. Differentiate in 70% ethanol.
4. 96% ethanol, 100% ethanol, toluene, Permount (Fisher).

For *myeloarchitectural* studies frozen sections of formalin-fixed material, cut at 100 µm, and stained after the following formula are recommended. The formula is a slight modification of the technique of Schroeder (1939).

1. Place sections in mordant solution for 48 h at 37°C.
   Keep sections in darkness.
   Prepare mordant by adding
           10.0 g potassium bichromate,
            2.5 g fluorchrom (Chroma),
            2.0 g sodium sulphate
      to 300.0 ml aqua dest.
   Boil shortly, let solution cool and filtrate.
2. Rinse shortly in aqua dest. (twice).
3. Stain in a solution of hematoxilin for 24 h at 37°C.
   Stock solution A: 10.0 g hematoxilin
                            2.0 g sodium iodate
                            100.0 ml 96% ethanol
   Stock solution B: Saturated aqueous solution of lithium carbonate
   Prepare staining solution by adding 3 ml of solution A to 100 ml aqua dest., boil for 5 min, let cool, and add 3 ml of solution B.
4. Rinse in aqua dest.
5. Oxidize in a 0.25% aqueous solution of potassium permanganate for about 1 min.

6. Rinse in aqua dest.
7. Bleach for about 1 min in a freshly prepared solution of
    1.0 g oxalic acid
    1.0 g potassium disulfite
    200.0 ml aqua dest.
8. Rinse in aqua dest. Repeat point 5 to 8 until background staining is eliminated.
9. Transfer to an aqueous solution of lithium carbonate (15 min). For preparation add 1 ml of solution B to 100 ml aqua dest.
10. Rinse in aqua dest. for several hours.
11. Transfer through a graded series of ethanol. In order to ensure flatness place sections between leaves of filter paper and press gently from above by means of for instance an exsiccator plate.
12. Transfer to toluene and Permount (Fisher).

For *pigmentoarchitectural* studies frozen sections of formalin-fixed material, cut at 800 μm, and stained after the following formula (H. Braak, 178e) are recommended.

1. Cut sections at 800 μm with the aid of a carbon dioxide-freezing microtome. Warm a block of aluminium with plane basis to about 35°C by putting it on a hot-plate of the kind used for the floating out and flattening of paraffin sections. Place warmed metal block on the surface of the frozen tissue for a few seconds. The time required depends on the size of both the metal block and the tissue block and can easily be found after a few trials. Sections should be cut at the dew-line, i.e., the border between the already thawed and the still frozen parts of the block. During sectioning the metal block is rewarmed on the hot-plate. A plano-concave knife, the kind used for celloidin sectioning, is recommended.
2. Place sections into a 4% aqueous solution of formaldehyde for about one week. Keep sections gently in movement by means of a shaking apparatus.
3. Rinse under running tap-water.
4. Oxidise in freshly prepared performic acid for about 30 min. Keep sections in movement. Prepare performic acid by adding 10 ml perhydrol (30% H_2O_2) to 90 ml 98%–100% formic acid. Let ripen for about 1 h, use solution for only one day (chemical hood! Handle agressive performic acid with care!).
5. Rinse under running tap-water for about 1 h.
6. Transfer to 70% ethanol (10 min).
7. Stain in a solution of aldehydefuchsin for about 12 h. Keep sections in movement.

Prepare stock solution by dissolving
- 0.5 g pararosanilin (Chroma)
- in 100.0 ml 70% ethanol
- add 1.0 ml 25% HCl, wait until clear and
- add 1.0 ml 100% crotonaldehyde.

Shake shortly and let ripen for about a week at room temperature. Use ripened stock solution for about one week only.
Prepare staining solution by adding 6 ml stock solution to 400 ml 96% ethanol plus 100 ml aqua dest. plus 50 ml 100% formic acid plus 5 ml performic acid (preparation see point 4). Prepare solution a day before staining and filtrate before use. Keep sections gently in movement by means of a shaking apparatus.

8. Rinse in 70% ethanol for about 10 min.
9. Dehydrate through a graded series of ethanol.
   In order to attain flatness place sections between pieces of filter paper and press gently during dehydration.
   Dehydrate for at least 24 h in 100% ethanol.
10. Transfer through toluene (several hours, use chemical hood) into Permount (Fisher). Addition of 10 ml 1-bromo-naphthalene to 90 ml Permount (Fisher) is recommended.

References

Ades HW, Felder R (1942) The acoustic area of the monkey (*Macaca mulatta*). J Neurophysiol 5: 49–54

Allmann JM, Kaas JH (1971a) A representation of the visual field in the caudal third of the middle temporal gyrus of the owl monkey (*Aotus trivirgatus*). Brain Res 31: 85–105

Allmann JM, Kaas JH (1971b) Representation of the visual field in striate and adjoining cortex of the owl monkey (*Aotus trivirgatus*). Brain Res 35: 89–106

Allmann JM, Kaas JH (1974) A crescent-shaped cortical visual area surrounding the middle temporal area (MT) in the owl monkey (*Aotus trivirgatus*). Brain Res 81: 199–213

Altschul R (1933) Die Glomeruli der Area praesubicularis. Z Gesamte Neurol Psychiatr 148: 50–54

Amaral DG (1978) A Golgi study of cell types in the hilar region of the hippocampus in the rat. J Comp Neurol 182: 851–914

Andersen P, Bliss TVP, Skrede KK (1971) Lamellar organisation of hippocampal excitatory pathways. Exp Brain Res 13: 222–238

Arseni C, Botez MI (1961) Speech disturbances caused by tumours of the supplementary motor area. Acta Psychiatr Neurol Scand 36: 279–299

Asanuma H, Rosén I (1972) Topographical organization of cortical efferent zones projecting to distal forelimb muscles in the monkey. Exp Brain Res 14: 243–256

Asanuma H, Ward JE (1971) Patterns of contraction of distal forelimb muscles produced by intracortical stimulation in cats. Brain Res 27: 97–109

Bailey P, Bonin G von (1951) The isocortex of man. University of Illinois Press, Urbana

Balthasar K (1954) Lebensgeschichte der vier größten Pyramidenzellarten in der V. Schicht der menschlichen Area gigantopyramidalis. J Hirnforsch 1: 281–325

Batsch G (1956) Die myeloarchitektonische Untergliederung des Isocortex parietalis beim Menschen. J Hirnforsch 2: 225–270

Bechterew W von (1891) Zur Frage über die äusseren Associationsfasern der Hirnrinde. Neurol Zentrbl 10: 682–684

Beck E (1934) Der Occipitallappen des Affen (*Macacus rhesus*) und des Menschen in seiner cytoarchitektonischen Struktur. I. *Macacus rhesus*. J Psychol Neurol 46: 193–323

Beck E (1936) Sensorische Aphasien. Z Neurol 158: 193–203

Benson DF (1967) Fluency in aphasia: Correlation with radioactive scan localization. Cortex 3: 373–394

Benson DF, Patten DH (1967) The use of radioactive isotopes in the localization of aphasia producing lesions. Cortex 3: 258–271

Berlin R (1858) Beitrag zur Structurlehre der Grosshirnwindungen. Inauguraldissertation. Junge, Erlangen

Betz W (1874) Anatomischer Nachweis zweier Gehirncentra. Zentralbl Med Wiss 12: 578–580, 594–599

Betz W (1881) Über die feinere Struktur der Gehirnrinde des Menschen. Zentralbl Med Wiss 19: 193–195, 210–213, 231–234

Billings-Gagliardi S, Chan-Palay V, Palay SL (1974) A review of lamination in area 17 of the visual cortex of *Macaca mulatta*. J Neurocytol 3: 619–629

References

Bishop GH, Smith JM (1964) The sizes of nerve fibers supplying cerebral cortex. Exp Neurol 9: 483–501

Blackstad TW, Kjaerheim A (1961) Special axo-dendritic synapses in the hippocampal cortex: Electron and light microscopic studies on the layer of mossy fibers. J Comp Neurol 117: 133–159

Blackstad TW, Brink K, Hem J, Jeune B (1970) Distribution of hippocampal mossy fibers in the rats. An experimental study with silver impregnation methods. J Comp Neurol 138: 433–450

Bohringer RC, Rowe MJ (1977) The organization of the sensory and motor areas of cerebral cortex in the Platypus (*Ornithorhynchus anatinus*). J Comp. Neurol 174: 1–14

Bondareff W, McLone DG (1973) The external glial limiting membrane in Macaca: Ultrastructure of a laminated glioepithelium. Am J Anat 136: 277–296

Bonin G von (1942) The striate area of primates. J Comp Neurol 77: 405–429

Bonin G von (1949) Architecture of the precentral motor cortex and some adjacent areas. In: Bucy PC (ed) The Precentral Motor Cortex, 2nd ed. The University of Illinois Press, Urbana, Illinois, pp 7–82

Bonin G von, Bailey P (1961) Pattern of the cerebral isocortex. In: Hofer H, Schulz AH, Starck D (eds) Primatologia II/2. Karger, Basel, New York

Bonin G von, Garol HW, McCulloch WS (1942) The functional organization of the occipital lobe. Biol Symp 7: 165–192

Braak E (1975) On the fine structure of the external glial layer in the isocortex of man. Cell Tissue Res 157: 367–390

Braak E (1976) On the fine structure of the small, heavily pigmented non-pyramidal cells in lamina II and upper lamina III of the human isocortex. Cell Tissue Res 169: 233–245

Braak E (1978a) On the structure of the human striate area. Lamina IVcβ. Cell Tissue Res. 188: 217–234

Braak E (1978b) Licht- und elektronenmikroskopische Untersuchungen zur Morphologie der primären Sehrinde des Menschen. Habilitationsschrift, Universität Kiel

Braak E (1980) On the structure of lamina IIIab pyramidal cells in the human isocortex. A Golgi and electron-microscopical study with special emphasis on the proximal axonal segment. J Hirnforsch 21: 439–444

Braak E, Drenckhahn D, Unsicker K, Gröschel-Stewart U, Dahl D (1978) Distribution of myosin and the glial fibrillary acidic protein (GFA protein) in rat spinal cord and in the human frontal cortex as revealed by immunofluorescence microscopy. Cell Tissue Res 191: 493–499

Braak E, Braak H, Strenge H, Muhtaroglu U (1980) Age-related alterations of the proximal axonal segment in lamina IIIab pyramidal cells of the human isocortex. A Golgi and fine structural study. J Hirnforsch 21 (in press)

Braak H (1971) Über das Neurolipofuscin in der unteren Olive und dem Nucleus dentatus cerebelli im Gehirn des Menschen. Z Zellforsch 121: 573–592

Braak H (1972a) Zur Pigmentarchitektonik der Großhirnrinde des Menschen. I. Regio entorhinalis. Z Zellforsch 127: 407–438

Braak H (1972b) Zur Pigmentarchitektonik der Großhirnrinde des Menschen. II. Subiculum. Z Zellforsch 131: 235–254

Braak H (1972c) Über die Kerngebiete des menschlichen Hirnstammes. V. Das dorsale Glossopharyngeus- und Vagusgebiet. Z. Zellforsch 135: 415–438

Braak H (1974a) On the structure of the human archicortex. I. The cornu ammonis. A Golgi and pigmentarchitectonic study. Cell Tissue Res 152: 349–383

Braak H (1974b) On pigment-loaded stellate cells within layer II and III of the human isocortex. A Golgi and pigmentarchitectonic study. Cell Tissue Res 155: 91–104

Braak H (1976a) A primitive gigantopyramidal field buried in the depth of the cingulate sulcus of the human brain. Brain Res 109: 219–233

Braak H (1976b) On the striate area of the human isocortex. A Golgi and pigmentarchitectonic study. J Comp Neurol 166: 341–364
Braak H (1977) The pigment architecture of the human occipital lobe. Anat. Embryol 150: 229–250
Braak H (1978a) On the pigmentarchitectonics of the human telencephalic cortex. In: Brazier MAB, Petsche H (eds) Architectonics of the cerebral cortex. Raven Press, New York, pp 137–157
Braak H (1978b) On magnopyramidal temporal fields in the human brain – probable morphological counterparts of Wernicke's sensory speech region. Anat Embryol 152: 141–169
Braak H (1978c) The pigment architecture of the human temporal lobe. Anat Embryol 154: 213–240
Braak H (1978d) Pigment architecture of the human telencephalic cortex. III. Regio praesubicularis. Cell Tissue Res 190: 509–523
Braak H (1978e) Eine ausführliche Beschreibung pigmentarchitektonischer Arbeitsverfahren. Mikroskopie 34: 215–221
Braak H (1979a) Spindle-shaped appendages of IIIab-pyramids filled with lipofuscin: A striking pathological change of the senescent human isocortex. Acta Neuropathol 46: 197–202
Braak H (1979b) The pigment architecture of the human frontal lobe. I. Precentral, subcentral and frontal region. Anat Embryol 157: 35–68
Braak H (1979c) Pigment architecture of the human telencephalic cortex. IV. Regio retrosplenialis. Cell Tissue Res 204: 431–440
Braak H (1979d) Pigment architecture of the human telencephalic cortex. V. Regio anterogenualis. Cell Tissue Res 204: 441–451
Braak H, Braak E (1976) The pyramidal cells of Betz within the cingulate and precentral gigantopyramidal field in the human brain. A Golgi and pigmentarchitectonic study. Cell Tissue Res 172: 103–119
Braak H, Braak E, Strenge H (1976) Gehören die Inselneurone der Regio entorhinalis zur Klasse der Pyramiden oder der Sternzellen? Z Mikrosk Anat Forsch 90: 1017–1031
Bradford R, Parnavelas JG, Lieberman AR (1977) Neurons in layer I of the developing occipital cortex of the rat. J Comp Neurol 176: 121–132
Braitenberg V (1962) A note on myeloarchitectonics. J Comp Neurol 118: 141–156
Braitenberg V (1974) Thoughts on the cerebral cortex. J Theoret Biol 46: 421–447
Braitenberg V (1978) Cortical architectonics: General and areal. In: Brazier MAB, Petsche H (eds) Architectonics of the cerebral cortex. Raven Press, New York, pp 443–465
Braitenberg V, Braitenberg C (1979) Geometry of orientation columns in the visual cortex. Biol Cybernetics 33: 179–186
Broca PP (1861) Perte de la parole, ramollissement chronique et destruction partielle du lobe antérieur gauche du cerveau. Bull Soc Anthrop (Paris) 2: 235–238
Broca PP (1863) Localisation des fonctions cérébrales. – Siége du langage articulé. Bull Soc Anthrop (Paris) 4: 200–204
Brockhaus H (1940) Die Cyto- und Myeloarchitektonik des Cortex claustralis und des Claustrum beim Menschen. J Psychol Neurol 49: 249–348
Brodmann K (1903) Beiträge zur histologischen Lokalisation der Grosshirnrinde. 1. Mitteilung: Die Regio Rolandica. J Psychol Neurol 2: 79–107
Brodmann K (1904) Beiträge zur histologischen Lokalisation der Grosshirnrinde. 2. Mitteilung: Der Calcarinatypus. J Psychol Neurol 2: 133–159
Brodmann K (1905/06) Beiträge zur histologischen Lokalisation der Grosshirnrinde. V. Mitteilung. Über den allgemeinen Bauplan des Cortex pallii bei den Mammaliern und zwei homologe Rindenfelder im besonderen. Zugleich ein Beitrag zur Furchenlehre. J Psychol Neurol 6: 275–400
Brodmann K (1908) Beiträge zur histologischen Lokalisation der Grosshirnrinde. VI. Mitteilung: Die Cortexgliederung des Menschen. J Psychol Neurol 10: 231–246

Brodmann K (1909) Vergleichende Lokalisationslehre der Grosshirnrinde in ihren Prinzipien dargestellt auf Grund des Zellenbaues. Barth, Leipzig

Brodmann K (1912) Neue Ergebnisse über die vergleichende histologische Lokalisation der Grosshirnrinde mit besonderer Berücksichtigung des Stirnhirns. Anat Anz Suppl 41: 157–216

Brodmann K (1914) Physiologie des Gehirns. In: Bruns P von (ed) Neue Deutsche Chirurgie, vol 11. Enke, Stuttgart, pp 85–426

Brown JW (1972) Aphasia, apraxia and agnosia. Clinical and theoretical aspects. Thomas, Springfield, Ill.

Bucy PC (1935) A comparative cytoarchitectonic study of the motor and premotor areas in the primate cortex. J Comp Neurol 62: 293–331

Burdach KF (1819–26) Vom Baue und Leben des Gehirns. Dyk, Leipzig

Butler AG, Jane JA (1977) Interlaminar connections of rat visual cortex: an ultrastructural study. J Comp Neurol 174: 521–534

Campbell AW (1905) Histological studies on the localization of cerebral function. University Press, Cambridge

Capanna E (1969) Considerazioni sul telencefalo degli anfibi. Atti Accad Naz Lincei Mem Ser 8: 55–82

Carrieri G (1963) Sindrome da sofferenza dell'area supplementare motoria sinistra nel corso di un meningioma parasagittale. Riv Patol Nerv Ment 84: 29–48

Casseday JH, Harting JK, Diamond JT (1976) Auditory pathways to the cortex in Tupaia glis. J Comp Neurol 166: 303–340

Celesia GG (1976) Organization of auditory cortical areas in man. Brain 99: 403–414

Chan-Palay V, Palay SL, Billings-Gagliardi SM (1974) Meynert cells in the primate visual cortex. J Neurocytol 3: 631–658

Chow KL (1961) Anatomical and electrophysiological analysis of temporal neocortex in relation to visual discrimination learning in monkeys. In: Delafresnaye JF (ed) Brain mechanisms and learning. Blackwell, Oxford, pp 507–525

Chow, KL, Pribram KH (1956) Cortical projection of the thalamic ventrolateral nuclear group in monkeys. J Comp Neurol 104: 57–75

Chronister RB, White LE (1975) Fiberarchitecture of the hippocampal formation: anatomy, projections, and structural significance. In: Isaacson RL, Pribram KH (eds) The hippocampus, vol I, Structure and development. Plenum Press, New York, London, pp 9–39

Chusid JG (1964) Black-and-white supplement for the color brain map of Bailey and von Bonin. Neurology 14: 154–157

Chusid JG, Guttierez-Mahoney CG de, Margulez-Lavergue MP (1954) Speech disturbances in association with parasagittal lesions. J Neurosurg 11: 193–204

Clairambault P, Capanna E (1970) Aspetti della neuroistogenesi del pallio telencefalico in due Rànidi. Rend Accad Naz Lincei Fasc 6 Ser 8, 18: 727–734

Clark WE LeGros (1942) The cells of Meynert in the visual cortex of the monkey. J Anat 76: 369–376

Clark WE LeGros, Powell, TPS (1953) On the thalamo-cortical connexions of the general sensory cortex of Macaca. Proc R Soc London Ser B 141: 467–487

Clark WE LeGros, Sunderland S (1939) Structural changes in the isolated visual cortex. J Anat 73: 563–574

Colonnier M (1967) The fine structural arrangement of the neocortex. Arch Neurol (Chicago) 16: 651–657

Conel JL (1939–1967) The postnatal development of the human cerebral cortex, vols I–VIII. Harvard Univ Press, Cambridge

Connolly C (1950) External morphology of the primate brain. Thomas, Springfield, Ill.

Cragg BG (1961) Olfactory and other afferent connections of the hippocampus in the rabbit, rat and cat. Exp Neurol 3: 588–600

Creswell GF, Reis DJ, MacLean PD (1964) Aldehyde-fuchsin positive material in brain of squirrel monkey (*Saimiri sciureus*). Am J Anat 115: 543–558

Crinis M de (1933) Über die Spezialzellen in der menschlichen Großhirnrinde. J Psychol Neurol 45: 439–449
Crinis M de (1934) Aufbau und Abbau der Großhirnleistungen und ihre anatomischen Gründe. Karger, Berlin
Cuénod M, Casey KL, MacLean PD (1965) Unit analysis of visual input to posterior limbic cortex. I. Phonic stimulation. J Neurophysiol 28: 1101–1117
Dean P, Cowey A (1977) Inferotemporal lesions and memory for pattern discrimination after visual interference. Neuropsychology 15: 93–98
Diamond IT (1967) The sensory neocortex. Contrib Sens Physiol 2: 51–100
Diamond IT (1976) Organization of the visual cortex: comparative anatomical and behavioral studies. Fed Proc 35: 60–67
Diamond IT, Hall WC (1969) Evolution of neocortex. Science 164: 251–262
Doedens KD, Nagel I, Schierhorn H (1974) Weitere Untersuchungen über die nackte (spinefreie) Initialzone der Apikaldendriten corticaler Pyramidenzellen der Albinoratte. Z Mikrosk Anat Forsch 88: 1093–1109
Doedens KD, Schierhorn H, Nagel I (1975) Die Ontogenese der spine-freien Initialzone der Apikaldendriten; Untersuchungen an corticalen Pyramidenzellen der Albinoratte. Gegenbaurs Morphol Jahrb 121: 88–108
Duckett S, Pearse AGE (1968) The cells of Cajal-Retzius in the developing human brain. J Anat 102: 183–187
Duvernoy H (1979) An angioarchitectonic study of the brain. Anat. Clin 1: 207–222
Eayrs JT, Goodhead B (1959) Postnatal development of the cerebral cortex in the rat. J Anat 93: 385–402
Economo C von (1926) Über den Zusammenhang der Gebilde des Retrosplenium. Z Zellforsch 3: 449–460
Economo C von (1927) Zellaufbau der Grosshirnrinde des Menschen. Zehn Vorlesungen. Springer, Berlin
Economo C von, Koskinas GN (1925) Die Cytoarchitektonik der Hirnrinde des erwachsenen Menschen. Springer, Wien Berlin
Englisch HJ, Kunz G, Wenzel J (1974) Zur Spines-Verteilung an Pyramidenneuronen der CA 1-Region des Hippocampus der Ratte. Z Mikrosk Anat Forsch 88: 85–102
Erickson TC, Woolsey CN (1951) Observations on the supplementary motor area of man. Trans Am Neurol Assoc 76: 50–56
Feldman M, Peters A (1978) The forms of non-pyramidal neurons in the visual cortex of the rat. J Comp Neurol 179: 761–794
Filimonoff IN (1932) Über die Variabilität der Großhirnrindenstruktur. II. Regio occipitalis beim erwachsenen Menschen. J Psychol Neurol 44: 1–96
Filimonoff IN (1933) Über die Variabilität der Großhirnrindenstruktur. III. Regio occipitalis bei den höheren und niederen Affen. J Psychol Neurol 45: 69–137
Filimonoff IN (1947) A rational subdivision of the cerebral cortex. Arch Neurol Psychiatry 58: 296–311
Fisken RA, Garey LJ, Powell TPS (1975) The intrinsic, association and commissural connections of area 17 of the visual cortex. Philos Trans R Soc London Ser B 272: 487–536
Flechsig P (1908) Bemerkungen über die Hörsphäre des menschlichen Gehirns. Neurol Zentralbl 27: 2–7, 50–57
Flechsig P (1920) Anatomie des menschlichen Gehirns und Rückenmarks auf myelogenetischer Grundlage, vol 1. Thieme, Leipzig
Flechsig P (1927) Meine myelogenetische Hirnlehre. Springer, Berlin
Fleischhauer K (1958) Darstellung rhinencephaler Strukturen durch Dithizon. Anat Anz 104: 135–137
Fleischhauer K (1959) Zur Chemoarchitektonik der Ammonsformation. Nervenarzt 30: 305–309
Fleischhauer K, Horstmann E (1957) Intravitale Dithizonfärbung homologer Felder der Ammonsformation fon Säugern. Z Zellforsch 46: 598–609

Fleischhauer K, Laube A (1979) Supracellular patterns in the cerebral cortex. In: Speckmann EJ, Caspers H (eds) Origin of cerebral field potentials. Thieme, Stuttgart, pp 1–12
Foerster O (1936) Motorische Felder und Bahnen. Sensible corticale Felder. In: Bumke O, Foerster O (eds) Handbuch der Neurologie, vol 6. Springer, Berlin, pp 1–448
Forbes BF, Moskowitz N (1974) Projections of auditory responsive cortex in the squirrel monkey. Brain Res 67: 239–254
Fox MW, Inman OR (1966) Persistence of Retzius-Cajal cells in the developing dog brain. Brain Res 3: 192–194
Fox MW, Inman OR, Himwich WA (1966) The postnatal development of neocortical neurons in the dog. J Comp Neurol 127: 199–206
Friede RL (1966) The histochemical architecture of the ammon'shorn as related to its selective vulnerability. Acta Neuropathol 6: 1–13
Frimmel G, Ost HM, Wenzel J (1975) Quantitative Untersuchungen zur Neuronenstruktur der Fascia dentata der Ratte. Z Mikrosk Anat Forsch 89: 495–511
Frotscher M (1975) Die postnatale Entwicklung corticaler Neurone und ihre Beeinflussung durch ein Trauma bei Rattus norvegicus B. J Hirnforsch 16: 203–221
Frotscher M, Mannsfeld B, Wenzel J (1975) Umweltabhängige Differenzierung der Dendritenspines an Pyramidenneuronen des Hippocampus (CA1) der Ratte. J Hirnforsch 16: 443–450
Frotscher M, Scharmacher K, Scharmacher M (1978a) Zur umweltabhängigen Differenzierung von Pyramidenneuronen im Hippocampus (CA1) der Ratte. Die Differenzierung von apikalen Seitendendriten und Basaldendriten. J Hirnforsch 19: 445–456
Frotscher M, Lößner B, Wenzel J (1978b) Quantitativ-elektronenmikroskopische Untersuchungen zur umweltabhängigen Differenzierung von Synapsen im Hippocampus (CA1) der Ratte. Z Mikrosk Anat Forsch 92: 171–182
Gaarskjaer FB (1978) Organization of the mossy fiber system of the rat studied in extended hippocampi. II. Experimental analysis of fiber distribution with silver impregnation methods. J Comp Neurol 178: 73–88
Garey LJ (1971) A light and electron microscopic study of the visual cortex of the cat and monkey. Proc R Soc London Ser B 179: 21–40
Garey LJ, Powell TPS (1971) An experimental study of the termination of the lateral geniculo-cortical pathway in the cat and monkey. Proc R Soc London Ser B 179: 41–63
Gatter KC, Powell TPS (1978) The intrinsic connections of the cortex of area 4 of the monkey. Brain 101: 513–541
Gennari F (1782) De peculiari structura cerebri. Parma
Gerhardt E (1940) Die Cytoarchitektonik des Isocortex parietalis beim Menschen. J Psychol Neurol 49: 367–419
Geschwind N (1965) Disconnection syndrome in animals and man. Brain 88: 237–294
Geschwind N (1974) Selected papers on language and the brain. Boston Studies, vol XVI. Reidel, Dordrecht, Boston
Geschwind N, Galaburda A, Le May M (1979) Morphological and physiological substrates of language and cognitive development. In: Katzman R (ed) Congenital and acquired cognitive disorders. Raven Press, New York, pp 31–41
Gihr M (1968) Klassifikation der großen Nervenzellen in der V. Rindenschicht der Area gigantopyramidalis des Menschen. J Hirnforsch 10: 413–439
Glickstein M, Whitteridge D (1974) Degeneration of layer III pyramidal cells in area 18 following destruction of callosal input. Anat Rec 178: 362–363
Globus A, Scheibel AB (1967) Pattern and field in cortical structure: The rabbit. J Comp Neurol 131: 155–172
Görne R, Pfister C (1976) Morphometrische Untersuchungen an praepyramidalen und Pyramiden-Neuronen verschiedener Vertebraten. Z Mikrosk Anat Forsch 90: 527–539

Gray EG (1959) Axosomatic and axo-dendritic synapses of the cerebral cortex: an electron microscopic study. J Anat 93: 420–434

Guidetti B (1957) Désordres de la parole associés a des lésions de la surface interhémisphérique frontale postérieure. Rev Neurol 97: 121–131

Hamlyn LH (1962) The fine structure of the mossy fibre endings in the hippocampus of the rabbit. J Anat. 96: 112–120

Hammarberg C (1898) Studien über Klinik und Pathologie der Idiotie nebst Untersuchungen über die normale Anatomie der Hirnrinde. Nova Acta Regiae Soc Sci Ups III/17: 1–126

Harrison JM, Howe ME (1974) Anatomy of the afferent auditory nervous system of mammals. In: Keidel WD, Neff WD (eds) Handbook of Sensory Physiology, vol V/1, Auditory system, anatomy, physiology (ear). Springer, Berlin Heidelberg New York, pp 283–336

Hassler R (1962) Die Entwicklung der Architektonik seit Brodmann und ihre Bedeutung für die moderne Hirnforschung. Dtsch Med Wochenschr 87: 1180–1185

Hassler R (1964) Zur funktionellen Anatomie des limbischen Systems. Nervenarzt 35: 386–396

Hassler R (1967) Funktionelle Neuroanatomie und Psychiatrie. In: Gruhle HW, Jung R, Mayer-Gross W, Müller M (eds) Psychiatrie der Gegenwart. Forschung und Praxis, vol I/1. Springer, Berlin Heidelberg New York, pp 152–285

Hassler R, Wagner A (1965) Experimentelle und morphologische Befunde über die vierfache corticale Projektion des visuellen Systems. 8th Int Congr Neurol 3: 77–96

Haug FMS (1967) Electron microscopical localization of the zinc in hippocampal mossy fibre synapses by a modified sulfide silver procedure. Histochemie 8: 355–368

Haug FMS, Blackstad TW, Simonsen AH, Zimmer J (1971) Timm's sulfide silver reaction for zinc during experimental anterograde degeneration of hippocampal mossy fibers. J Comp Neurol 142: 23–32

Haug H (1971) Die Membrana limitans gliae superficialis der Sehrinde der Katze. Z Zellforsch 115: 79–87

Haug H, Kölln M, Rast A (1976) The postnatal development of myelinated nerve fibres in the visual cortex of the cat. A stereological and electron microscopical investigation. Cell Tissue Res 167: 265–288

Hécaen H, Consoli S (1973) Analyse des troubles du langage au cours des lesions de l'aire de Broca. Neuropsychologia 11: 377–388

Hécaen H, Penfield W, Bertrand C, Malmo R (1956) The syndrome of apractognosia due to lesions of the minor cerebral hemisphere. Arch Neurol Psychiatry 75: 400–434

Hendrickson AR, Wilson JR, Ogren MP (1978) The neuroanatomical organization of pathways between the dorsal lateral geniculate nucleus and visual cortex in old world and new world primates. J Comp Neurol 182: 123–136

Hind JE, Benjamin RM, Woolsey CN (1958) Auditory cortex of squirrel monkey (*Saimiri sciureus*). Fed Proc 17: 71

Hjorth-Simonsen A (1972) Projection of the lateral part of the entorhinal area to the hippocampus and fascia dentata. J Comp Neurol 146: 219–231

Hoesen GW van, Pandya DN (1973) Afferent and efferent connections of the perirhinal cortex (area 35) in the rhesus monkey. Anat Rec 175: 460–461

Hoesen GW van, Pandya DN, Butters N (1972) Cortical afferents to the entorhinal cortex of the rhesus monkey. Science 175: 1471–1473

Holloway RL (1968) The evolution of the primate brain: Some aspects of quantitative relations. Brain Res 7: 121–172

Hopf A (1954) Die Myeloarchitektonik des Isocortex temporalis beim Menschen. J Hirnforsch 1: 208–279

Hopf A (1955) Über die Verteilung myeloarchitektonischer Merkmale in der isokortikalen Schläfenlappenrinde beim Menschen. J Hirnforsch 2: 36–54

Hopf A (1956) Über die Verteilung myeloarchitektonischer Merkmale in der Stirnhirnrinde beim Menschen. J Hirnforsch 2: 311–333
Hopf A (1957) Architektonische Untersuchungen an sensorischen Aphasien. J Hirnforsch 3: 275–530
Hopf A (1968) Photometric studies on the myeloarchitecture of the human temporal lobe. J Hirnforsch 10: 285–297
Hopf A (1969/70) Photometric studies on the myeloarchitecture of the human parietal lobe. I. Parietal region. J Hirnforsch 11: 253–265
Hopf A (1970) Photometric studies on the myeloarchitecture of the human parietal lobe. II. Postcentral region. J Hirnforsch 12: 135–141
Hopf A, Vitzthum H Gräfin (1957) Über die Verteilung myeloarchitektonischer Merkmale in der Scheitellappenrinde beim Menschen. J Hirnforsch 3: 79–104
Hubel DH, Wiesel TN (1972) Laminar and columnar distribution of geniculo-cortico fibers in the macaque monkey. J Comp Neurol 146: 421–450
Hubel DH, Wiesel TN (1977) Ferrier lecture. Functional architecture of macaque monkey visual cortex. Proc R Soc London Ser B 198: 1–59
Hyvärinen J, Poranen A (1974) Function of the parietal associative area 7 as revealed from cellular discharges in alert monkeys. Brain 97: 673–692
Imig TJ, Ruggero MA, Kitzes LM, Javel E, Brugge JF (1977) Organization of auditory cortex in the owl monkey (*Aotus trivirgatus*). J Comp Neurol 171: 111–128
Ingvar DH (1975) Patterns of brain activity revealed by measurements of regional cerebral blood flow. In: Ingvar DH, Lassen NA (eds) Alfred Benzon Symposium VIII: Brain work. Munksgaard Publ, Copenhagen, pp 397–413
Ingvar DH (1976) Functional landscapes of the dominant hemisphere. Brain Res 107: 181–197
Ingvar DH (1978) Localisation of cortical functions by multiregional measurements of the cerebral blood flow. In: Brazier MAB, Petsche H (eds) Architectonics of the cerebral cortex. Raven Press, New York, pp 235–243
Ingvar DH, Schwartz MS (1974) Blood flow patterns induced in the dominant hemisphere by speech and reading. Brain 97: 273–288
Ingvar DH, Philipson L (1977) Distribution of cerebral blood flow in the dominant hemisphere during motor ideation and motor performance. Ann Neurol 2: 230–237
Iversen SD (1970) Interference and inferotemporal memory deficits. Brain Res 19: 277–289
Jacobson M (1969) Development of specific neuronal connections. Science 163: 543–547
Jacobson M (1970a) Development, specification and diversification of neuronal connections. In: Schmitt FO (ed) The neurosciences. Second Study Programme. Rockefeller Press, New York, pp 116–129
Jacobson M (1970b) Developmental neurobiology. Holt, Rinehart and Winston, New York
Jacobson M (1974) A plenitude of neurons. In: Gottlieb G (ed) Studies on the development of behavior and the nervous system, vol II, Aspects of neurogenesis. Academic Press, London New York, pp 151–166
Jacobson M (1975) Development and evolution of type II neurons: conjectures a century after Golgi. In: Santini M (ed) Golgi Centenial Symp Proc. Raven Press, New York, pp 147–151
Jacobson S (1963) Sequence of myelinization in the brain of the albino rat. A. Cerebral cortex, thalamus and related structures. J Comp Neurol 121: 5–29
Jacobson S (1965) Intralaminar, interlaminar, callosal, and thalamo-cortical connection in frontal and parietal areas of the albino rat cerebral cortex. J Comp Neurol 124: 131–146
Jacobson S (1967) Dimensions of the dendritic spine in the sensorimotor cortex of the rat, cat, squirrel monkey and man. J Comp Neurol 129: 49–58

Janzen RWC (1967) Topographische Besonderheiten im Bau der Glia marginalis des Menschen. Z Zellforsch 80: 570–584

Jones EG (1975) Varieties and distribution of non-pyramidal cells in the somatic sensory cortex of the squirrel monkey. J Comp Neurol 160: 205–268

Jones EG, Powell TPS (1969a) Synapses on the axon hillocks and initial segments of pyramidal cell axon in the cerebral cortex. J Cell Sci 5: 495–507

Jones EG, Powell TPS (1969b) Morphological variations in the dendritic spines of the neocortex. J Cell Sci 5: 509–529

Jones EG, Powell TPS (1970a) Electron microscopy of the somatic sensory cortex of the cat. I. Cell types and synaptic organization. Philos Trans R Soc London Ser B 257: 1–11

Jones EG, Powell TPS (1970b) Electron microscopy of the somatic sensory cortex of the cat. II. The fine structure of layers I and II. Philos Trans R Soc. London Ser B 257: 13–21

Jones EG, Powell TPS (1970c) Electron microscopy of the somatic sensory cortex of the cat. III. The fine structure of layers III to VI. Philos Trans R Soc London SerB 257: 23–28

Jones EG, Powell TPS (1970d) Connexions of the somatic sensory cortex of the rhesus monkey. III. Thalamic connexions. Brain 93: 37–56

Jones EG, Powell TPS (1970e) An anatomical study of converging sensory pathways within the cerebral cortex of the monkey. Brain 93: 793–820

Jones EG, Burton H (1976) Areal differences in the laminar distribution of thalamic afferents in cortical fields of the insular, parietal and temporal regions of primates. J Comp Neurol 168: 197–247

Jones EG, Coulter JD, Hendry SHC (1978) Intracortical connectivity of architectonic fields in the somatic sensory, motor and parietal cortex of monkeys. J Comp Neurol 181: 291–384

Kaas JH, Hall WC, Killackey H, Diamond IT (1972) Visual cortex of the tree shrew (*Tupaia glis*): architectonic subdivisions and representations of the visual field. Brain Res 42: 491–496

Kaas JH, Nelson RJ, Sur M, Lin CS, Merzenich MM (1979) Multiple representations of the body within the primary somatosensory cortex of primates. Science 204: 521–523

Kaes T (1907) Die Großhirnrinde des Menschen in ihren Maßen und in ihrem Fasergehalt. Fischer, Jena

Kaiserman-Abramof JR, Peters A (1972) Some aspects of the morphology of Betz cells in the cerebral cortex of the cat. Brain Res 43: 527–546

Kasdon DL, Jacobson S (1978) The thalamic afferents to the inferior parietal lobule of the rhesus monkey. J Comp Neurol 177: 685–706

Kawata A (1927) Zur Myeloarchitektonik der menschlichen Hirnrinde. Arb Neurol Inst Wien Univ 29: 191–225

Kertesz A, Lesk D, McCabe P (1977) Isotope localization of infarcts in aphasia. Arch Neurol (Chicago) 34: 590–601

Kirsche W (1972) Die Entwicklung des Telencephalons der Reptilien und deren Beziehung zur Hirn-Bauplanlehre. Nova Acta Leopoldina 37/2 (NF 204). Barth, Leipzig

Kirsche W (1974) Zur vergleichenden funktionsbezogenen Morphologie der Hirnrinde der Wirbeltiere auf der Grundlage embryologischer und neurohistologischer Untersuchungen. Z Mikrosk Anat Forsch 88: 21–51

Kirsche W, Kirsche K (1962) Zur Fibrilloarchitektonik des Neocortex von *Macaca mulatta* Zimmermann. J Hirnforsch 5: 83–125

Kirsche W, Kunz G, Wenzel J, Wenzel M, Winkelmann A, Winkelmann E (1973) Neurohistologische Untersuchungen zur Variabilität der Pyramidenzellen des sensomotorischen Cortex der Ratte. J Hirnforsch 14: 117–135

Klingler J (1948) Die makroskopische Anatomie der Ammonsformation. Denkschriften der Schweizerischen Naturforschenden Gesellschaft, vol 78. Fretz, Zürich

Koelliker A von (1896) Handbuch der Gewebelehre des Menschen, vol II. Nervensystem des Menschen und der Thiere, 6th ed. Engelmann, Leipzig

Koelliker A von (1899) Sind der Spitzenbesatz der Dendriten der Neurodendren normale Bildungen oder ein Kunstprodukt? In: Erinnerungen aus meinem Leben. Engelmann, Leipzig, pp 241–246

König N, Valat J, Fulcrand J, Marty R (1977) The time of origin of Cajal-Retzius cells in the rat temporal cortex. An autoradiographic study. Neurosci Lett 4: 21–26

Kretschmann HJ, Schleicher A, Grottschreiber JF, Kullmann W (1979) The Yakovlev collection – A pilot study of its suitability for the morphometric documentation of the human brain. J Neurol Sci 43: 111–126

Krettek JE, Price JL (1977a) Projections from the amygdaloid complex to the cerebral cortex and thalamus in the rat and cat. J Comp Neurol 172: 687–722

Krettek JE, Price JL (1977b) Projections from the amygdaloid complex and adjacent olfactory structures to the entorhinal cortex and to the subiculum in the rat and cat. J Comp Neurol 172: 723–752

Kruska D, Stephan H (1973) Volumenvergleich allokortikaler Hirnzentren bei Wild- und Hausschweinen. Acta Anat 84: 387–415

Kunz G, Kirsche W, Wenzel J, Winckelmann E, Neumann H (1972) Quantitative Untersuchungen über die Dendritenspines an Pyramidenneuronen des sensorischen Cortex der Ratte. Z Mikrosk Anat Forsch 85: 397–416

Kunz G, Holz L, Englisch HJ, Wenzel J (1974) Untersuchungen zur Spinedichte und Spineverteilung an apikalen und basalen Dendriten von Lamina-V-Pyramiden-Zellen des sensorischen Cortex der Ratte. J Hirnforsch 15: 389–399

Kunz G, Englisch HJ, Wenzel J (1976) Untersuchungen der Spines-Verteilung an Pyramidenneuronen der CA_1-Region des Hippocampus der Ratte nach langzeitiger oraler Alkoholapplikation. J Hirnforsch 17: 351–364

Laatsch RH, Cowan WM (1966) Electron microscopic studies of the dentate gyrus of the rat. I. Normal structure with special reference to synaptic organization. J. Comp Neurol 128: 359–395

Laplane D, Talairach J, Meininger V, Bancaud J, Orgogozo, JM (1977) Clinical consequences of corticectomies involving the supplementary motor area in man. J Neurol Sci 34: 301–314

Larsen B, Skinhoj E, Lassen NA (1978) Variations in regional cortical blood flow in the right and left hemispheres during automatic speech. Brain 101: 193–209

Lassek AM (1940) The human pyramidal tract. II. A numerical investigation of the Betz cells of the motor area. Arch Neurol Psychiatry 42: 872–876

Lassen NA, Ingvar DH, Skinhoj E (1978a) Brain function and blood flow. Sci Am 239: 50–59

Lassen NA, Larsen B, Orgogozo JM (1978b) Les localisations corticales vues par la gamma-caméra dynamique: une nouvelle approche en neuropsychologie. Encephale 4: 233–249

Lende RA (1963) Cerebral cortex: A sensorimotor amalgam in the marsupialia. Science 141: 730–732

Lende RA, Sadler KM (1967) Sensory motor areas in neocortex of hedgehog (*Erinaceus*). Brain Res 5: 390–405

Le Vay S (1973) Synaptic patterns in the visual cortex of the cat and monkey. Electron microscopy of Golgi preparations. J Comp Neurol 150: 53–86

Levey NH, Jane JA (1975) Laminar thermocoagulation of the visual cortex of the rat. I. Interlaminar connections. Brain Behav Evol 11: 257–274

Lindsay RD, Scheibel AB (1976) Quantitative analysis of dendritic branching pattern of granular cells from human dentate gyrus. Exp Neurol 52: 295–310

Locke S (1961) The projection of the magnocellular medial geniculate body. J Comp Neurol 116: 179–193 (1961)

Loos H van der (1965) The "improperly" oriented pyramidal cell in the cerebral cortex and its possible bearing on problems of neuronal growth and cell orientation. Bull Johns Hopkins Hosp 117: 228–250

Lopes CAS, Mair WGP (1974) Ultrastructure of the outer cortex and the pia mater in man. Acta Neuropathol 28: 79–86

Lorente de Nó R (1933) Studies on the structure of the cerebral cortex. I. The area entorhinalis. J Psychol Neurol 45: 381–438

Lorente de Nó R (1934) Studies on the structure of the cerebral cortex. II. Continuation of the study of the ammonic system. J Psychol Neurol 46: 113–177

Lorente de Nó R (1938) The cerebral cortex: Architecture, intracortical connections and motor projections. In: Fulton JF (ed) Physiology of the nervous system. Oxford, London New York Toronto, pp 291–321

Lund JS (1973) Organisation of neurons in the visual cortex, area 17, of the monkey (*Macaca mulatta*). J Comp Neurol 147: 455–496

Lund JS, Boothe RG (1975) Interlaminar connections and pyramidal neuron organisation in the visual cortex, area 17, of the macaque monkey. J Comp Neurol 159: 305–334

Lund JS, Boothe RG, Lund RD (1977) Development of neurons in the visual cortex (area 17) of the monkey (*Macaca nemestrina*). A Golgi study from fetal day 127 to postnatal maturity. J Comp Neurol 176: 149–188

Lungwitz W (1937) Zur myeloarchitektonischen Untergliederung der menschlichen Area praeoccipitalis (Area 19 Brodmann). J Psychol Neurol 47: 607–638

Lynch G, Smith RL, Mensah P, Cotmann C (1973) Tracing the dentate gyrus mossy fiber system with horseradish peroxidase histochemistry. Exp Neurol 40: 516–524

MacLean PD, Creswell G (1970) Anatomical connections of visual system with limbic cortex of monkey. J Comp Neurol 138: 265–278

Mann, DMA, Yates PO (1974) Lipoprotein pigments – their relationship to ageing in the human nervous system. I. The lipofuscin content of nerve cells. Brain 97: 481–488

Mannen H (1955) La cytoarchitecture du système nerveux central humain regardée au point de vue de la distribution de grains de pigments jaunes contenant de la graisse, Acta Anat Nipp 30: 151–174

Marin-Padilla M (1970a) Prenatal and early postnatal ontogenesis of the human motor cortex: a Golgi study. I. The sequential development of the cortical layers. Brain Res 23: 167–183

Marin-Padilla M (1970b) Prenatal and early postnatal ontogenesis of the human motor cortex. II. The basket-pyramidal system. Brain Res 23: 185–191

Marin-Padilla M (1971) Early prenatal ontogenesis of the cerebral cortex (neocortex) of the cat (*Felis domestica*). A Golgi study. I. The primordial neocortical organisation. Z Anat Entwicklungsgesch 134: 117–145

Marin-Padilla M (1972) Prenatal ontogenetic history of the principal neurons of the neocortex of the cat (*Felis domestica*). A Golgi study. II. Developmental differences and their significances. Z Anat Entwicklungsgesch 136: 125–142

Marin-Padilla M, Stibitz GR, Almy CP, Brown HN (1969) Spine distribution of the layer V pyramidal cell in man. Brain Res 12: 493–496

Martinotti C (1890) Beitrag zum Studium der Hirnrinde und dem Centralursprung der Nerven. Int Monatsschr Anat Physiol 7: 69–90

Masdeu JC, Schoene WC, Funkenstein H (1978) Aphasia following infarction of the left supplementary motor area. Neurology 28: 1220–1223

Maske H (1955) Über den topochemischen Nachweis von Zink im Ammonshorn verschiedener Säugetiere. Naturwissenschaften 42: 424

Massopust LC, Wolin LR, Kadoya S (1968) Evoked responses in the auditory cortex of the squirrel monkey. Exp Neurol 21: 35–40

Maurer J, Fleischhauer K (1979) Preferential orientation of small profiles in the neuropil of lamina I. Anat Embryol 157: 133–149

Meller K, Breipohl W, Glees P (1968a) The cytology of the developing molecular layer of mouse motor cortex. An electron microscopical and a Golgi impregnation study. Z Zellforsch 86: 171–183

Meller K, Breipohl W, Glees P (1968b) Synaptic organization of the molecular and the outer granular layer in the motor cortex in the white mouse during postnatal development. A Golgi- and electronmicroscopical study. Z Zellforsch 92: 217–231

Meller K, Breipohl W, Glees P (1969) Ontogeny of the mouse motor cortex. The polymorph layer or layer VI. A Golgi and electronmicroscopical study. Z Zellforsch 99: 443–458

Merzenich MM, Brugge JF (1973) Representation of the cochlear partition on the superior temporal plane of the macaque monkey. Brain Res 50: 275–296

Merzenich MM, Knight PL, Roth GL (1973) Cochleotopic organization of primary auditory cortex in the cat. Brain Res 63: 343–346

Merzenich MM, Kaas JH, Sur M, Lin C (1978) Double representation of the body surface within cytoarchitectonic areas 3b and 1 in "S1" in the owl monkey (*Aotus trivirgatus*). J Comp Neurol 181: 41–74

Mesulam MM, Pandya DN (1973) The projections of the medial geniculate complex within the Sylvian fissure of the rhesus monkey. Brain Res 60: 315–334

Mesulam MM, Hoesen GW van, Pandya DN, Geschwind N (1977) Limbic and sensory connections of the inferior parietal lobule (area PG) in the rhesus monkey. A study with a new method for horseradish peroxidase histochemistry. Brain Res 136: 393–414

Mettler FA (1949) Cytoarchitecture. In: Mettler FA (ed) Selective partial ablation of the frontal cortex. A correlative study of its effects on human psychotic subjects. Hoeber, New York, pp 48–78

Meyer A (1977) The search for a morphological substrate in the brains of eminent persons including musicians: a historical review. In: Critchley M, Henson RA (eds) Music and the brain. Heinemann, London

Meynert T (1868) Der Bau der Gross-Hirnrinde und seine örtlichen Verschiedenheiten, nebst einem pathologisch-anatomischen Corollarium. Vierteljahresschr Psychiatr 2: 88–113

Meynert T (1872) Vom Gehirne der Säugethiere. In: Stricker S (ed) Handbuch der Lehre von den Geweben des Menschen, vol 2. Engelmann, Leipzig, pp 694–808

Michalski A, Patzwaldt R, Schulz E, Schönheit B (1976) Quantitative Untersuchungen an primitiven Pyramidenzellen in der vorderen cingulären Rinde der Ratte. J Hirnforsch 17: 143–153

Minkwitz HG (1976a) Zur Entwicklung der Neuronenstruktur des Hippocampus während der prä- und postnatalen Ontogenese der Albinoratte. I. Mitteilung: Neurohistologische Darstellung der Entwicklung langaxoniger Neurone aus den Regionen CA_3 und CA_4. J Hirnforsch 17: 213–231

Minkwitz HG (1976b) Zur Entwicklung der Neuronenstruktur des Hippocampus während der prä- und postnatalen Ontogenese der Albinoratte. II. Mitteilung: Neurohistologische Darstellung der Entwicklung von Interneuronen und des Zusammenhanges lang- und kurzaxoniger Neurone. J Hirnforsch 17: 233–253

Minkwitz HG (1976c) Zur Entwicklung der Neuronenstruktur des Hippocampus während der prä- und postnatalen Ontogenese der Albinoratte. III. Mitteilung: Morphometrische Erfassung der ontogenetischen Veränderungen in Dendritenstruktur und Spinebesatz an Pyramidenneuronen (CA_1) des Hippocampus. J Hirnforsch 17: 255–275

Mishkin M (1972) Cortical visual areas and their interaction. In: Karzsmar AG, Eccles JC (eds) The brain and human behavior. Springer, Berlin Heidelberg New York, pp 187–208

Mitra NL (1955) Quantitative analysis of cell types in mammalian neocortex. J Anat 89: 467–483

Moffie D (1949) The parietal lobe. A survey of its anatomy and functions. Folia Psychiatr Neurol Neurochir Neerl 52: 418–444

Molliver ME, Loos H van der (1969/70) The ontogenesis of cortical circuitry: the spatial distribution of synapses in somesthetic cortex of newborn dog. Ergebn Anat Entwicklungsgesch 42: 1–53

Morest DK (1969) The growth of dendrites in the mammalian brain. Z Anat Entwicklungsgesch 128: 290–317
Myers RE (1962) Commissural connections between occipital lobes of the monkey. J Comp Neurol 118: 1–16
Myers RE (1965) Organization of visual pathways. In: Ettlinger EG (ed) Functions of the corpus callosum. Churchill, London, pp 133–138
Nañagas JC (1923) Anatomical studies on the motor cortex of macacus rhesus. J Comp Neurol 35: 67–96
Nauta HJW, Butler AB, Jane JA (1973) Some observations on axonal degeneration resulting from superficial lesion of the cerebral cortex. J Comp Neurol 150: 349–360
Ngowyang G (1932) Beschreibung einer Art von Spezialzellen in der Inselrinde, zugleich Bemerkungen über die v. Economoschen Spezialzellen. J Psychol Neurol 44: 671–674
Niessing K (1936) Über systemartige Zusammenhänge der Neuroglia im Großhirn und über ihre funktionelle Bedeutung. Gegenbaurs Morphol Jahrb 78: 537–584
Niklowitz W, Bak IJ (1965) Elektronenmikroskopische Untersuchungen am Ammonshorn. I. Die normale Substruktur der Pyramidenzellen. Z Zellforsch 66: 529–547
Noback CR, Purpura DP (1961) Postnatal ontogenesis of neurons in cat neocortex. J Comp Neurol 117: 291–308
Northcutt RG (1967) Architectonic studies of the telencephalon of *Iguana iguana*. J Comp Neurol 130: 109–147
Obersteiner H (1903) Über das hellgelbe Pigment in den Nervenzellen und das Vorkommen weiterer fettähnlicher Körper im Centralnervensystem. Arb Neurol Inst Wien 10: 245–274
Obersteiner H (1904) Weitere Bemerkungen über die Fett-Pigment-Körnchen im Centralnervensystem. Arb Neurol Inst Wien 11: 400–406
Ojemann GA, Whitaker HA (1978) Language localization and variability. Brain, Language 6: 239–260
Oliver DL, Hall WC (1978) The medial geniculate body of the tree shrew, *Tupaia glis*. II. Connections with the neocortex. J Comp Neurol 182: 459–494
Oppermann K (1929) Cajalsche Horizontalzellen und Ganglienzellen des Markes. Z Neurol Psychiatr 120: 121–137
Otsuka N, Kawamoto M (1966) Histochemische und autoradiographische Untersuchungen der Hippocampusformation der Maus. Histochemie 6: 267–273
Palay SL (1978) The Meynert cell, an unusual cortical pyramidal cell. In: Brazier MAB, Petsche H (eds) Architectonics of the cerebral cortex. Raven Press, New York, pp 31–42
Palay SL, Sotelo C, Peters A, Orkand PM (1968) The axon hillock and the initial segment. J Cell Biol 38: 193–201
Pandya DN, Sanidès F (1973) Architectonic parcellation of the temporal operculum in rhesus monkey and its projection pattern. Z Anat Entwicklungsgesch 139: 127–161
Penfield W (1966) Speech, perception and the uncommitted cortex. In: Eccles JC (ed) Brain and conscious experience. Springer, Berlin Heidelberg New York, pp 217–237
Penfield W, Boldrey E (1937) Somatic motor and sensory representation in the cerebral cortex of man as studied by electrical stimulation. Brain 60: 389–443
Penfield W, Rasmussen T (1949) Vocalization and arrest of speech. Arch Neurol Psychiatry 61: 21–27
Penfield W, Rasmussen T (1950) The cerebral cortex of man. MacMillan, New York
Penfield W, Roberts L (1959) Speech and brain-mechanisms. University Press, Princeton
Peters A (1971) Stellate cells of the rat parietal cortex. J Comp Neurol 141: 345–374
Peters A, Fairén A (1978) Smooth and sparsely-spined stellate cells in the visual cortex of the rat: A study using a combined Golgi-electron microscope technique. J Comp Neurol 181: 129–172
Peters A, Kaiserman-Abramof IR (1969) The small pyramidal neuron of the cerebral cortex. The synapses upon dendritic spines. Z Zellforsch 100: 487–506

Peters A, Kaiserman-Abramof IR (1970) The small pyramidal neuron of the rat cerebral cortex. The perikaryon, dendrites and spines. Am J Anat 127: 321–356

Peters A, Proskauer CC, Kaiserman-Abramof IR (1968) The small pyramidal neuron of the rat cerebral cortex. The axon hillock and initial segment. J Cell Biol 39: 604–619

Peters A, Palay SL, Webster HDF (1970) The fine structure of the nervous system. The cell and their processes. Harper and Row, New York

Petit-Dutaillis D, Giot G, Messimy R, Bourdillon C (1954) A propos d'une aphémie par atteinte de la zone motrice supplémentaire de Penfield, au cours de l'évolution d'un anévrism artério-veineux. Guérison de l'aphémie par ablation de la lésion. Rev Neurol 90: 95–106

Petrides M, Iversen SD (1979) Restricted posterior parietal lesions in the rhesus monkey and performance on visuospatial tasks. Brain Res 161: 63–77

Pfeifer RA (1920) Myelogenetisch-anatomische Untersuchungen über das kortikale Ende der Hörleitung. Abh Sächs Akad Wiss Math Phys Kl 37: 1–54

Pfeifer RA (1936) Pathologie der Hörstrahlung und der corticalen Hörsphäre. In: Bumke O, Foerster O (eds) Handbuch der Neurologie, vol VI. Springer, Berlin, pp 533–626

Pfeifer RA (1940) Die angioarchitektonische areale Gliederung der Großhirnrinde. Thieme, Leipzig

Poliakov GI (1961) Some results of research into the development of the neuronal structure of the cortical ends of the analyzers in man. J Comp Neurol 117: 197–212

Poliakov GI (1964/65) Development and complication of the cortical part of the coupling mechanism in the evolution of vertebrates. J Hirnforsch 7: 253–273

Poliakov GI (1966) Embryonal and postembryonal development of neurons of the human cerebral cortex. In: Hassler R, Stephan H (eds) Evolution of the forebrain. Phylogenesis and ontogenesis of the forebrain. Thieme, Stuttgart, pp 249–258

Poliakov GI (1972/73) Some morphological aspects of the integrating activity of the cerebral cortex. J Hirnforsch 13: 469–487

Polyak S (1957) The vertebrate visual system. University of Chicago Press, Chicago

Powell TPS, Mountcastle VB (1959a) The cytoarchitecture of the postcentral gyrus of the monkey *macaca mulatta*. Bull Johns Hopkins Hosp 105: 108–131

Powell TPS, Mountcastle VB (1959b) Some aspects of functional organization of the cortex of the postcentral gyrus of the monkey: a correlation of findings obtained in a single unit analysis with cytoarchitecture. Bull Johns Hopkins Hosp 105: 133–162

Powell TPS, Cowan WM, Raisman G (1965) The central olfactory connexions. J Anat 99: 791–813

Price JL (1973) An autoradiographic study of complementary laminar patterns of termination of afferent fibers to the olfactory cortex. J Comp Neurol 150: 87–108

Price JL, Powell TPS (1971) Certain observations on the olfactory pathway. J Anat 110: 105–126

Pubols BH, Pubols LM (1971) Somatotopic organisation of spider monkey somatic sensory cerebral cortex. J Comp Neurol 141: 63–76

Purpura DP (1975) Morphogenesis of visual cortex in the preterm infant. In: Brazier MAB (ed) Growth and development of the brain. Raven Press, New York, pp 33–49

Purpura DP, Housepian EM (1961) Morphological and physiological properties of chronically isolated immature neocortex. Exp Neurol 4: 377–401

Purpura DP, Shofer RJ, Housepian EM, Noback CR (1964) Comparative ontogenesis of structure-function relations in cerebral and cerebellar cortex. In: Purpura DP, Schadé JP (eds) Growth and maturation of the brain. Prog Brain Res 4: 187–221

Raedler A, Sievers J (1975) The development of the visual system of the albino rat. Adv Anat Embryol Cell Biol 50: fasc 3

Raichle ME, Grubb RL, Mokhtar HG, Eichling JO, Ter-Pogossian MM (1976) Correlation between regional cerebral blood flow and oxidative metabolism. Arch Neurol (Chicago) 33: 523–526

Rakic P (1975) Effects of local cellular environments on the differentiation of LCN's. Neurosci Res Progr Bull 13: 400–407

Ramón y Cajal S (1891) Sur la structure de l'écorce cérébrale de quelques mammifères Cellule 7: 123–172

Ramón y Cajal S (1893) Estructura del asta de ammon. Anal Soc Esp Hist Nat Madrid 22, in German: Beiträge zur feineren Anatomie des großen Hirns. I. Über die feinere Struktur des Ammonshornes. Z Wiss Zool 56: 615–663

Ramón y Cajal S (1900a) Studien über die Hirnrinde des Menschen. 1. Heft: Die Sehrinde. Barth, Leipzig

Ramón y Cajal S (1900b) Studien über die Hirnrinde des Menschen. 2. Heft: Die Bewegungsrinde. Barth, Leipzig

Ramón y Cajal S (1902) Studien über die Hirnrinde des Menschen. 3. Heft: Die Hörrinde. Barth, Leipzig

Ramón y Cajal S (1903) Studien über die Hirnrinde des Menschen. 4. Heft: Die Riechrinde beim Menschen und Säugetier. Barth, Leipzig

Ramón y Cajal S (1906) Studien über die Hirnrinde des Menschen. Heft 5: Vergleichende Strukturbeschreibung und Histogenesis der Hirnrinde. Barth, Leipzig

Ramón y Cajal S (1909) Histologie du système nerveux de l'homme et des vertébrés. Maloine, Paris (Madrid: Consejo superior de investigaciones científicas. Reprinted 1952 and 1955)

Ramón-Moliner E (1961) The histology of the postcruciate gyrus in the cat. III. Further observations. J Comp Neurol 117: 229–249

Ramsey HJ (1965) Fine structure of the surface of the cerebral cortex of human brain. J Cell Biol 26: 323–333

Rasmussen T, Milner B (1975) Clinical and surgical studies of the cerebral speech areas in man. In: Zülch KJ, Creutzfeldt O, Galbraith GC (eds) Cerebral localization. Springer, Berlin Heidelberg New York, pp 238–257

Retzius G (1893) Die Cajal'schen Zellen der Großhirnrinde beim Menschen und bei Säugetieren. Biol Unters 5: 1–9

Retzius G (1894) Weitere Beiträge zur Kenntnis der Cajal'schen Zellen der Großhirnrinde des Menschen. Biol Unters 6: 29–34

Ribak CE (1978) Aspinous and sparsely-spinous stellate neurons in the visual cortex of rats contain glutamic acid decarboxylase. J Neurocytol 7: 401–478

Rickmann M, Chronwall BM, Wolff IR (1977) On the development of non-pyramidal neurons and axons outside the cortical plate: The early marginal zone as a pallial anlage. Anat. Embryol 151: 285–307

Risberg J, Ingvar DH (1973) Patterns of activation in the grey matter of the dominant hemisphere during memoring and reasoning. Brain 76: 737–756

Roland PE, Larsen B (1976) Focal increase of cerebral blood flow during stereognostic testing in man. Arch Neurol 33: 551–558

Roland PE, Larsen B, Lassen NA, Skinhøj E (1980) Supplementary motor area and other cortical areas in organization of voluntary movements in man. J. Neurophysiol 43: 118–136

Rosabal F (1967) Cytoarchitecture of the frontal lobe of the squirrel monkey. J Comp Neurol 130: 87–108

Rose J (1938) Zur normalen und pathologischen Architektonik der Ammonsformation. I. Normalanatomischer Teil. J Psychol Neurol 49: 137–188

Rose M (1927a) Der Allocortex bei Tier und Mensch. I. Teil. J Psychol Neurol 34: 1–111

Rose M (1927b) Die sog. Riechrinde beim Menschen und beim Affen. II. Teil des „Allocortex bei Tier und Mensch". J Psychol Neurol 34: 261–401

Rose M (1928) Gyrus limbicus anterior and Regio retrosplenialis (Cortex holoprotoptychos quinquestratificatus). Vergleichende Architektonik bei Tier und Mensch. J Psychol Neurol 35: 65–173

Rose M (1935) Cytoarchitektonik und Myeloarchitektonik der Großhirnrinde. In: Bumke O, Foerster O (eds) Handbuch der Neurologie, vol 1. Springer, Berlin, pp 588–778

Rose S (1927) Vergleichende Messungen im Allocortex bei Tier und Mensch. J Psychol Neurol 34: 250–255

Rosene DL, Hoesen GW van (1977) Hippocampal efferents reach widespread areas of cerebral cortex and amygdala in the rhesus monkey. Science 198: 315–317

Rubens AB (1975) Aphasia with infarction in the territory of the anterior cerebral artery. Cortex 11, 239–250

Russell WR, Espir MLE (1961) Traumatic aphasia. A study of aphasia in war wounds of the brain. Oxford University Press, London

Rutledge LT (1978) Effects of cortical denervation and stimulation on axons, dendrites and synapses. In: Cotman CW (ed) Neuronal plasticity. Raven Press, New York, pp 273–289

Sala L (1891) Zur feineren Anatomie des großen Seepferdfußes. Z Wiss Zool 52: 18–45

Sanides D, Sanides F (1974) A comparative Golgi study of the neocortex in insectivores and rodents. Z Mikrosk Anat Forsch 88: 959–977

Sanides F (1962) Die Architektonik des menschlichen Stirnhirns. In: Müller M, Spatz H, Vogel P (eds) Monographien aus dem Gesamtgebiete der Neurologie und Psychiatrie, vol 98. Springer, Berlin Göttingen Heidelberg

Sanides F (1963) Die Architektonik des menschlichen Stirnhirns und die Prinzipien seiner Entwicklung. Fortschr Med 81: 831–838

Sanides F (1964) The cyto-myeloarchitecture of the human frontal lobe and its relation to phylogenetic differentiation of the cerebral cortex. J Hirnforsch 6: 269–282

Sanides F (1968) The architecture of the cortical taste nerve areas in squirrel monkey (*Saimiri sciureus*) and their relationships to insular, sensorimotor and prefrontal regions. Brain Res 8: 97–124

Sanides F (1969) Comparative architectonics of the neocortex of mammals and their evolutionary interpretation. Ann NY Acad Sci 167: 404–423

Sanides F (1970) Functional architecture of motor and sensory cortices in primates in the light of a new concept of neocortex evolution. In: Noback C, Montagna W (eds) Advances in primatology, vol I: The primate brain. Meredith, New York, pp 137–208

Sanides F (1971) Evolutionary aspect of the primate neocortex. Proc 3rd Int Congr Primat, Zürich 1970, vol 1. Karger, Basel, pp 92–98

Sanides F (1972) Representation in the cerebral cortex and its areal lamination patterns. In: Bourne GA (ed) Structure and function of nervous tissue, vol 1. Academic Press, London New York, pp 329–453

Sanides F (1975) Comparative neurology of the temporal lobe in primates including man with reference to speech. Brain Language 2: 396–419

Sanides F, Gräfin Vitzthum H (1965a) Zur Architektonik der menschlichen Sehrinde und den Prinzipien ihrer Entwicklung. Dtsch Z Nervenheilk 187: 680–707

Sanides F, Gräfin Vitzthum H (1965b) Die Grenzerscheinungen am Rande der menschlichen Sehrinde. Dtsch Z Nervenheilk 187: 708–719

Sanides F, Hoffmann J (1969) Cyto- and myeloarchitecture of the visual cortex of the cat and of the surrounding integration cortices. J Hirnforsch 11: 79–104

Sanides F, Krishnamurti A (1967) Cytoarchitectonic subdivisions of sensorimotor and prefrontal regions and of bordering insular and limbic fields in slow loris (*Nycticebus coucang coucang*). J Hirnforsch 9: 225–252

Sanides F, Sas E (1970) Persistence of horizontal cells of the Cajal fetal type and of the subpial granular layer in parts of the mammalian palaeocortex. Z Mikrosk Anat Forsch 82: 570–588

Sanides F, Sanides D (1972) The 'extraverted neurons' of the mammalian cerebral cortex. Z Anat Entwicklungsgesch 136: 272–293

Sarkissov SA, Filimonoff IN, Kononowa EP, Preobraschenskaja IS, Kukuew LA (1955) Atlas of the cytoarchitectonics of the human cerebral cortex. Medgiz, Moscow

Sas E, Sanides F (1970) A comparative Golgi study of Cajal foetal cells. Z Mikrosk Anat Forsch 82: 385–396

Schadé JP, Groenigen WB van (1961) Structural organization of the human cerebral cortex. 1. Maturation of the middle frontal gyrus. Acta Anat 47: 74–111

Schadé JP, Backer H van, Colon E (1964a) Quantitative analysis of neuronal parameters in the maturing cerebral cortex. Prog Brain Res 4: 150–175

Schadé JP, Meeter K, Groenigen WB van (1964b) Maturational aspects of the dendrites in the human cerebral cortex. Acta Morphol Neerl Scand 5: 37–48

Schaffer K (1892) Beitrag zur Histologie der Ammonshornformation. Arch Mikrosk Anat 39: 611–632

Scheibel ME, Scheibel AB (1978) The dendritic structure of the human Betz cell. In: Brazier MAB, Petsche H (eds) Architectonics of the cerebral cortex. Raven Press, New York, pp 43–57

Scheibel ME, Davies TL, Lindsay RD, Scheibel AB (1974) Basilar dendrite bundles of giant pyramidal cells. Exp Neurol 42: 307–319

Schierhorn H (1978a) Die postnatale Entwicklung der Lamina V-Pyramidenzellen im sensomotorischen Cortex der Albinoratte. 1. Einleitung und qualitative Untersuchung von Golgi-Präparaten. Gegenbaurs Morphol Jahrb 124: 1–23

Schierhorn H (1978b) Die postnatale Entwicklung der Lamina V-Pyramidenzellen im sensomotorischen Cortex der Albinoratte. 2. Quantitative Untersuchung von Golgi-Präparaten. Gegenbaurs Morphol Jahrb 124: 24–42

Schierhorn H (1978c) Die postnatale Entwicklung der Lamina V-Pyramidenzellen im sensomotorischen Cortex der Albinoratte. 3. Diskussion und Schrifttum. Gegenbaurs Morphol Jahrb 124: 230–255

Schierhorn H, Bossanyi P von, Nagel I, Weber T (1972a) Spine-Verteilung an den Apikaldendriten der großen Pyramidenzellen (Lamina V) in der sensomotorischen und limbischen Hirnrinde der Ratte. Gegenbaurs Morphol Jahrb 118: 423–447

Schierhorn H, Doedens K, Nagel I (1972b) Über die laminäre Zuordnung der apikalen Dendritenspines der Lamina-V-Pyramiden in der sensomotorischen Hirnrinde der Albinoratte. Gegenbaurs Morphol Jahrb 118: 465–487

Schierhorn H, Doedens K, Nagel I (1973) Über das spine-freie („nackte") Intitialsegment der Apikaldendriten von corticalen Pyramidenzellen der Albinoratte. Gegenbaurs Morphol Jahrb 119: 130–145

Schiller F (1947) Aphasia studied in patients with missile wounds. J Neurol Neurosurg Psychiatry 10: 183–197

Schleicher A, Zilles K, Wingert F, Kretschmann HJ (1975) Bestimmungen der Anzahl der Zellen mit mehr als einem Nucleolus im histologischen Schnittpräparat. Microsc Acta 77: 316–330

Schönheit B, Schulz E (1976) Quantitative Untersuchungen über die Dendriten-Spines an den Lamina V-Pyramidenzellen im Bereich der vorderen cingulären Rinde der Ratte. J Hirnforsch 17: 171–187

Schroeder K (1939) Eine weitere Verbesserung meiner Markscheidenfärbemethode am Gefrierschnitt. Z Gesamte Neurol 166: 588–593

Schulz E, Schönheit B (1974) Neurohistologische Untersuchungen zur Neuronenstruktur der Regio limbica anterior der Ratte. J Hirnforsch 15: 469–490

Schulz E, Schönheit B, Holz L (1976) Quantitative Untersuchungen am Dendritenbaum von großen (regulären) Pyramidenzellen der Lamina V im Bereich der vorderen cingulären Rinde der Ratte. J Hirnforsch 17: 155–168

Schwerdtfeger WK (1979) Direct efferent and afferent connections of the hippocampus with the neocortex in the marmoset monkey. Am J Anat 156: 77–82

References

Segal M, Landis S (1974) Afferents to the hippocampus of the rat studied with the method of retrograde transport of horseradish peroxidase. Brain Res 78: 1–15

Seltzer B, Pandya DN (1974) Polysensory cortical projections to the parahippocampal gyrus in the rhesus monkey. Anat Rec 178: 460–461

Sgonina K (1937) Vergleichende Anatomie der Entorhinal- und Präsubikularregion. J Psychol Neurol 48: 56–163

Shkol'nik-Yarros EG (1971) Neurons and interneuronal connections of the central visual system. Plenum Press, New York

Shoumura K, Ando T, Kato K (1975) Structural organization of 'callosal' OBg in human callosum agenesis. Brain Res 93: 241–252

Shute CCD, Lewis PR (1967) The ascending cholinergic reticular system: neocortical, olfactory and subcortical projections. Brain 90: 497–520

Smit GJ, Uylings HBM (1975) The morphometry of the branching pattern in dendrites of the visual cortex pyramidal cells. Brain Res 87: 41–54

Smit GJ, Uylings HBM, Veldmaat-Wansink L (1972) The branching pattern in dendrites of cortical neurons. Acta Morphol Neerl Scand 9: 253–274

Smith GE (1907) A new topographical survey of the human cerebral cortex, being an account of the distribution of the anatomically distinct cortical areas and their relationship to the cerebral sulci. J Anat 41: 237–254

Solcher H (1958) Zur Cytologie und Cytoarchitektonik des Präzentralgebietes. Dtsch Z Nervenheilkd 178: 89–95

Solnitzky O, Harman PJ (1946) The regio occipitalis of the lorisiform lemuroid *Galago demidovii*. J Comp Neurol 84: 339–384

Spatz WB, Tigges J, Tigges M (1970) Subcortical projections, cortical associations and some intrinsic interlaminar connections of the striate cortex in the squirrel monkey (*Saimiri*). J Comp Neurol 140: 155–174

Stephan H (1954) Vergleichend-anatomische Untersuchungen an Hirnen von Wild- und Haustieren. III. Die Oberflächen des Allocortex von Wild- und Gefangenschaftsfüchsen. Biol Zentralbl 73: 96–115

Stephan H (1956) Vergleichend-anatomische Untersuchungen an Insektivorengehirnen. II. Oberflächenmessungen am Allocortex im Hinblick auf funktionelle und phylogenetische Probleme. Gegenbaurs Morphol Jahrb 97: 123–142

Stephan H (1960) Die quantitative Zusammensetzung der Oberflächen des Allocortex bei Insektivoren und Primaten. In: Tower DB, Schadé IP (eds) Structure and function of the cerebral cortex. Elsevier, Amsterdam London New York Princetown, pp 51–58

Stephan H (1961) Vergleichend-anatomische Untersuchungen an Insektivorengehirnen. V. Die quantitative Zusammensetzung der Oberflächen des Allocortex. Acta Anat 44: 12–59

Stephan H (1963) Vergleichend-anatomische Untersuchungen am Uncus bei Insektivoren und Primaten. In: The rhinencephalon and related structures. Prog Brain Res 3: 111–121

Stephan H (1964) Die kortikalen Anteile des limbischen Systems (Morphologie und Entwicklung). Nervenarzt 35: 396–401

Stephan H (1966) Größenänderungen im olfaktorischen und limbischen System während der phylogenetischen Entwicklung der Primaten. In: Hassler R, Stephan H (eds) Evolution of the forebrain. Thieme, Stuttgart, pp 377–388

Stephan H (1975) Allocortex. In: Bargmann W (ed) Handbuch der mikroskopischen Anatomie des Menschen, vol IV/9. Springer, Berlin Heidelberg New York

Stephan H, Andy OJ (1970) The allocortex in primates. In: Noback CR, Montagna W (eds) The primate brain. Advances in primatology, vol I. Appleton-Century-Crofts, New York, pp 109–135

Strasburger EH (1937) Die myeloarchitektonische Gliederung des Stirnhirns beim Menschen und Schimpansen. J Psychol Neurol 47: 461–491

Talairach J, Bancaud J (1966) The supplementary motor area in man (anatomo-functional findings by stereo-electroencephalography in epilepsy). Int J Neurol 5: 330–347

Talairach J, Bancaud J, Geier S, Bordas-Ferrer M, Bonis A, Szikla G, Rusu M (1973) The cingulate gyrus and human behavior. Electroencephalogr Clin Neurophysiol 34: 45–52

Thomas H (1966) Licht- und elektronenmikroskopische Untersuchungen an den weichen Hirnhäuten und den Pacchionischen Granulationen des Menschen. Z Mikrosk Anat Forsch 75: 270–327

Tigges M, Bos J, Tigges J, Bridges E (1977) Ultrastructural characteristics of layer IV neuropil in area 17 of monkeys. Cell Tissue Res 182: 39–59

Timm F (1958) Zur Histochemie des Ammonshorngebietes. Z Zellforsch 48: 548–555

Tömböl T (1972) A Golgi analysis of the sensory-motor cortex in the rabbit. In: Petsche H, Brazier MAB (eds) Synchronization of EEG activity in epilepsies. Springer, Wien, pp 25–36

Tömböl T (1978) Comparative data on the Golgi architecture of interneurons of different cortical areas in cat and rabbit. In: Brazier MAB, Petsche H (eds) Architectonics of the cerebral cortex. Raven Press, New York, pp 59–76

Turowski A, Danner H (1977) Zur Morphologie des Telencephalons von *Salmo irideus* (Teleostei). Golgi-Imprägnationsstudie. J Hirnforsch 18: 37–51

Ulinski PS (1974) Cytoarchitecture of cerebral cortex in snakes. J Comp Neurol 158: 243–266

Uylings HBM, Smit GJ (1975) Three-dimensional branching structure of pyramidal cell dendrites. Brain Res 87: 55–60

Valverde F (1965) Studies on the piriform lobe. Harvard University Press, Cambridge (Mass)

Valverde F (1977) Lamination of the striate cortex. J Neurocytol 6: 483–484

Valverde F (1978) The organization of area 18 in the monkey. A Golgi study. Anat Embryol 154: 305–334

Vitzthum H Gräfin (1959) Die Architektonik der Rinde des Scheitellappens. Psychiatr Neurol Med Psychol 12: 325–331

Vitzthum H Gräfin, Sanides F (1966) Entwicklungsprinzipien der menschlichen Sehrinde. In: Hassler R, Stephan H (eds) Evolution of the forebrain. Thieme, Stuttgart, pp 435–442

Vogt BA (1976) Retrosplenial cortex in the rhesus monkey: a cytoarchitectonic and Golgi study. J Comp Neurol 169: 63–98

Vogt BA, Rosene DL, Pandya DN (1979) Thalamic and cortical afferents differentiate anterior from posterior cingulate cortex in the monkey. Science 204: 205–207

Vogt C, Vogt O (1919) Allgemeinere Ergebnisse unserer Hirnforschung. J Psychol Neurol 25: 279–461

Vogt C, Vogt O (1926) Die vergleichend-architektonische und die vergleichend-reizphysiologische Felderung der Großhirnrinde unter besonderer Berücksichtigung der menschlichen. Naturwissenschaften 14: 1190–1194

Vogt C, Vogt O (1937) Sitz und Wesen der Krankheiten im Lichte der topistischen Hirnforschung und des Variierens der Tiere. J Psychol Neurol 47: 237–457

Vogt C, Vogt O (1940) Das formative Sonderverhalten des einzelnen Griseum cerebrale. Forsch Fortschr 16: 274–276

Vogt C, Vogt O (1942) Morphologische Gestaltungen unter normalen und pathogenen Bedingungen. J Psychol Neurol 50: 161–524

Vogt C, Vogt O (1954) Gestaltung der topistischen Hirnforschung und ihre Förderung durch den Hirnbau und seine Anomalien. J Hirnforsch 1: 1–46

Vogt C, Vogt O (1956) Weitere Ausführungen zum Arbeitsprogramm des Hirnforschungsinstitutes in Neustadt (Schwarzwald). J Hirnforsch 2: 403–427

Vogt M (1928) Über omnilaminäre Strukturdifferenzen und lineare Grenzen der architektonischen Felder der hinteren Zentralwindung des Menschen. J Psychol Neurol 35: 177–193

Vogt M (1929) Über fokale Besonderheiten der Area occipitalis im cytoarchitektonischen Bilde. J Psychol Neurol 39: 506–510
Vogt O (1910) Die myeloarchitektonische Felderung des menschlichen Stirnhirns. J Psychol Neurol 15: 221–232
Vogt O (1911/12) Die Myeloarchitektonik des Isocortex parietalis. J Psychol Neurol 18: 379–390
Vogt O (1927) Architektonik der menschlichen Hirnrinde. Allg Z Psychiatr 86: 247–274
Vogt O (1941) Über nationale Hirnforschungsinstitute. J Psychol Neurol 50: 1–10
Wallace RB, Kaplan R, Werboff J (1977) Hippocampus and behavioral maturation. Intern J Neurosci 7: 185–200
Walsh TM, Ebner FF (1970) The cytoarchitecture of somatic sensory-motor cortex in the opossum (*Didelphis marsupialis virginiana*): a Golgi study. J Anat 107: 1–18
Walshe FMR (1942) The giant pyramidal cells of Betz, the motor cortex and the pyramidal tract: a critical review. Brain 65: 409–461
Wenzel J, Bogolepov NN (1976) Elektronenmikroskopische und morphometrische Untersuchungen zur Synaptologie des Hippocampus der Ratte. J Hirnforsch 17: 399–448
Wenzel J, Wenzel M, Kirsche W, Kunz G, Neumann H (1972) Quantitative Untersuchungen über die Verteilung der Dendritenspines an Pyramidenneuronen des Hippocampus (CA 1). Z Mikrosk Anat Forsch 85: 23–34
Wenzel J, Kirsche W, Kunz G, Neumann H, Wenzel M, Winkelmann E (1973) Licht- und elektronenmikroskopische Untersuchungen über die Dendritenspines an Pyramiden-Neuronen des Hippocampus (CA 1) bei der Ratte. J Hirnforsch 13: 387–408
Wenzel J, Kammerer E, Frotscher M, Joschko R, Joschko M, Kaufmann W (1977) Elektronenmikroskopische und morphometrische Untersuchungen an Synapsen des Hippocampus nach Lernexperimenten bei der Ratte. Z Mikrosk Anat Forsch 91: 74–93
Werner G, Whitsel BL (1971) The functional organization of the somatosensory cortex. In: Iggo A (ed) Handbook of Sensory Physiology. Springer, Berlin Heidelberg New York, pp 621–700
Wernicke C (1874) Der aphasische Symptomencomplex. Eine psychologische Studie auf anatomischer Basis. Cohn und Weigert, Breslau
West C (1979) A quantitative study of lipofuscin accumulation with age in normals and individuals with Down's syndrome, phenylketonuria, progeria and transneuronal atrophy. J Comp Neurol 186: 109–116
Whitaker HA, Selnes OA (1976) Anatomic variations in the cortex: Individual differences and the problem of the localization of language functions. Ann NY Acad Sci 280: 844–854
Whitsel BL, Rustioni A, Dreyer DA, Loe PR, Allen EE, Metz CB (1978) Thalamic projections to S-I in macaque monkey. J Comp Neurol 178: 385–410
Whitteridge D (1973) Projection of optic pathways to the visual cortex. In: Jung, R (ed) Handbook of Sensory Physiology, vol VIII/3 B. Springer, Berlin Heidelberg New York, pp 247–268
Williams RS, Ferrante RJ, Caviness VS (1978) The Golgi-rapid method in clinical neuropathology. I. Morphologic consequences of suboptimal fixation. J Neuropathol Exp Neurol 37: 13–33
Williams RS, Ferrante RJ, Caviness VS (1979) The isolated human cortex. A Golgi analysis of Krabbe's disease. Arch Neurol (Chicago) 36: 134–139
Woolsey CN (1955) Organization of somatic sensory and motor areas of the cerebral cortex. In: Harlow H, Woolsey CN (eds) Biological and biochemical bases of behavior. University of Wisconsin Press, Madison, pp 63–81
Woolsey CN (1961) Organisation of cortical auditory system. In: Rosenblith WA (ed) Sensory communication. Wiley and Sons, New York, pp 235–257

Woolsey CN, Fairman D (1946) Contralateral, ipsilateral, and bilateral representation of cutaneous receptors in somatic areas I and II of the cerebral cortex of pigs, sheep, and other mammals. Surgery 19: 684–702

Woolsey CN, Settlage PH, Meyer DR, Sencer W, Pinto Hamuy T, Travis AM (1952) Patterns of localization in precentral and "supplementary" motor areas and their relation to the concept of a premotor area. Res Publ Assoc Res Nerv Ment Dis 30: 238–264

Yakovlev PI, Lecours AR (1967) The myelogenetic cycles of regional maturation of the brain. In: Minkowski A (ed) Development of the brain in early life. Blackwell Sci. Publ., Oxford Edinburgh, pp 3–70

Yakovlev PI, Locke S, Koskoff DY, Patton RA (1960) Limbic nuclei of thalamus and connections of limbic cortex. I. Organization of the projections of the anterior group of nuclei and of the midline nuclei of the thalamus to the anterior cingulate gyrus and hippocampal rudiment in the monkey. Arch Neurol (Chicago) 3: 620–641

Zangwill OL (1975) Excision of Broca's area without persistent aphasia. In: Zülch KJ, Creutzfeld O, Galbraith GC (eds) Cerebral localization. Springer, Berlin Heidelberg New York, pp 258–263

Zeki SM (1969) The secondary visual areas of the monkey. Brain Res 13: 197–226

Zeki SM (1970) Interhemispheric connections of prestriate cortex in monkey. Brain Res 19: 63–75

Zeki SM (1977) Colour coding in the superior temporal sulcus of rhesus monkey visual cortex. Proc R Soc London Ser B 197: 195–223

Zilles K, Schleicher A, Kretschmann HJ, Wingert F (1976) Semiautomatic morphometric analysis of the nucleolar development in the nucl. n. oculomotorii of *Tupaia belangeri* during ontogenesis. Anat Embryol 149: 15–28

Zilles K, Schleicher A, Kretschmann HJ (1978) A quantitative approach to cytoarchitectonics. I. The areal pattern of the cortex of *Tupaia belangeri*. Anat Embryol 153: 195–212

Zilles K, Rehkämper G, Stephan H, Schleicher A (1979) A quantitative approach to cytoarchitectonics. IV. The areal pattern of the cortex of *Galago demidovii* (E. Geoffroy, 1796), (Lorisidae, Primates). Anat Embryol 157: 81–103

Subject Index

Primary discussions will be found on the pages given in *italics*

Allocortex 24, *26–48*
Alveus 28, 29
Angioarchitectonics 1
Anterogenual region 49, *55-61*
Area anterogenualis magnoganglionaris 58–60
– – simplex 58, *60–61*
– ectogenualis 58
– ectosplenialis 50
– frontalis dysganglionaris inf., sup. et paracentr. 91, 96–97
– – gigantoganglionaris 91, *94–96*
– – magnoganglionaris 91, 96
– – transganglionaris 91, 97
– inferofrontalis magnopyramidalis centralis 98–99
– paraganglionaris magnopyramidalis centralis 99
– paragenualis 61
– parasplenialis 54
– parastriata 64, *70–73*
– parietalis granulosa 85, 86–88
– – paragranulosa 85, 88
– peristriata magnopyramidalis 64, *73–74*
– retrosplenialis intermedia 50–52
– – lateralis 50–52
– – medialis 54
– striata *64–70*
– superofrontalis magnopyramidalis centralis 99–100
– temporalis granulosa 74, *76–78*
– – magna 84
– – magnopyramidalis centralis *80–82*
– – paragranulosa 76, 78
– – progranulosa 78
– – stratiformis 84, 85
Areae entorhinales 39–42
– parasubiculares 36
– presubiculares 36
– subiculares 31–33
– transentorhinales 42–44
– transsubiculares 37

Baillarger, lines of 13, 16–17, 19, 21–22, *55*
Basket cells 26, 29
Belt areas 63
– –, motor 61, 97–98
– –, sensory 54, 70–74, 78, 84, 88, 90
Betz cells 7, 14, *22*, 60, 88, *96–97*
Brain maps 104–120

Chemoarchitectonics 1
Coniocortex 15, 28, 34, 50–52, 64–70, 76–78, 86–88
Core fields 63, 77
– –, motor 30–31, 58–61, 93–97
– –, sensory 28, 34, 50–52, 64–70, 74, 76–78, 84
Corkscrew cells 60
Cornu ammonis *28–30*, 46, 49, 58
Cytoarchitectonic lamination 13–15
Cytoarchitectonics 1

Dendrogenesis 1
Domestication 37
Double-bush pyramidal cells 29

Entorhinal region 37–48
External glial layer 18, *66*
– tenia 13, 19, *54–55*
Extraverted neurons 9

Fascia dentata *26–28*, 46, 49
Fasciola cinerea 26

Gennari, line of *64*, 68, 69
Granule cells, fascia dentata 9, *26–28*
– –, isocortex 14–15
Gyrus ambiens 37, 42
– fasciolaris 28
– frontalis inferior 98
– – superior 99, 100
– parahippocampalis 35, 37
– postcentralis 88, 90
– precentralis 94, 97
– temporalis inferior 76, 84

Gyrus temporalis medius 76, 84
– – superior 76
– – transversus primus 76
– uncinatus 30, 34

Hippocampal formation 26–33
– –, supracallosal parts 49–50, 58
Horizontal cells of Cajal 11, 18

Induseum griseum 49, 58
Internal tenia 13, 19, 61
Isocortex 12, 24, 63–103

Kaes-Bechterew, line of 13, 16, 17, 21, 72, 73, 88

Lamina cellularis profunda 41
– dissecans 37, 41
– limitans gliae externa 18, 66
– tecti 63
Limbus Giacomini 26
Limes parastriatus 73
Lobulus paracentralis 93
– parietalis inferior 90

Magnopyramidal regions 44, 100
– –, allocortical 44
– –, frontal 91, 98–99
– –, occipital 64, 73–74
– –, parietal 85, 90
– –, temporal 76, 80–84
Martinotti cells 11
Mesoneocortex 25
Mesocortex 25
Meynert cells 6, 14, 69–70
Microdendrites 9, 29
Mossy fibres 28, 29, 30
Myeloarchitectonics 1, 15–17
Myelogenesis 1

Non-pyramidal cells 10–11

Operculum frontoparietalis 98
Organ-pipe formation 82

Pair of compasses cells 6
Parasubiculum 34, 36
Perforant path 48
Periallocortex 24, 25, 37, 42–44, 49–50
Periamygdalear region 44
Peristriate region 73–74
Pigment-laden IIIc-pyramids 19, 44, 73, 80, 90, 98–100
– Vb-pyramids 19

– VIa-neurons 97
– stellate cells 10–11, 18, 25, 80
Pigmentoarchitectonic lamination 13, 18–23
Pigmentoarchitectonics 1, 18–23
Prepiriform region 44
Presubiculum 34–37, 46
– proper 34, 36
Proisocortex 24, 25, 49–62
Properistriata 34
Pyramidal cells 3–6
– –, abnormally oriented 4
– –, axon 4
– –, axonal enlargements 19
– –, dendrites 3–4
– –, dendritic pigment 31
– –, dendritic spines 4, 5
– –, lipofuscin content 5
– –, modified 6–10, 29, 34, 68
– –, Nissl granules 5–6

Rainshower formation 77
Regio anterogenualis 55–62
– entorhinalis 44
– frontalis paraganglionaris 91, 97–99
– inferofrontalis magnopyramidalis 91, 98–99
– occipitalis magnopyramidalis 64, 73–74
– parietalis magnopyramidalis 85, 90
– periamygdalearis 44
– prepiriformis 44
– presubicularis 34–37
– retrosplenialis 49–55
– subcentralis 86, 97
– temporalis magnopyramidalis 76, 80–84
– temporalis uniteniata 76, 84–85
Regional cortical blood flow 100–101
Retrosplenial region 34, 49–55

Solitary cells of Cajal 7, 68
Sommer's sector 30
Speech centre of Broca 99–100
– – of Wernicke 82–84
Speech centres 100
Stellate cells 10–11
– –, axon 10
– –, dendrites 10
– –, double-bush variety 68–69
– –, lipofuscin content 10–11
– –, spiny variety 68–69
Stick cells 60
Stratum oriens 26, 29
Subiculum 30–33, 46, 50, 58
Substratum eumoleculare 30

Subject Index

Substratum lacunosum 30
– radiatum 29–30
Sulcus calcarinus 70
– centralis 86, 88, 93–94, 100
– diagonalis 98
– fimbriodentatus 26
– frontalis inferior 86, 88, 94, 97
– hippocampalis 26
– parieto-occipitalis 70

Tassel cells 41
Tenia externa 13, 19, *54–55*
– interna 13, 19, *61*
Transentorhinal subregion 42–44
Transsubiculum 34, 37

Uncus 26, 28, 35

Verrucae hippocampi 37

Wernicke centre 82–84

Studies of Brain Function

Coordinating Editor:
V. Braitenberg
Editors:
H. B. Barlow, E. Bizzi, F. Florey,
O.-J. Grüsser, H. van der Loos

Volume 1
W. Heiligenberg

Principles of Electrolocation and Jamming Avoidance in Electric Fish

A Neuroethological Approach

1977. 58 figures, 1 table. XI, 85 pages
ISBN 3-540-08367-7

Contents: General Physiological and Anatomical Background: The Electric Organ. Electroreceptors. Taxonomy of Electrolocating Fish. The Spectral Composition of Electric Organ Discharges. The Neuroanatomy of Electric Fish. – The Mechanism of Electrolocation: Spatial Aspects of Electrolocation. Response Characteristics and Central Projections of Tuberous Electroreceptors. Central Processing of Electric Images. Behavioral Measures of Electrolocation Performance. Electrolocation Performance in the Presence of Electric Noise and Mechanisms of Jamming Avoidance. Neuronal Mechanisms Linked to Jamming Avoidance and Electrolocation Under Jamming Conditions. Hypotheses and Results. Speculations on the Evolution of Pulse- and Wave-Type Electric Fish.

Volume 2
W. Precht

Neuronal Operations in the Vestibular System

1978. 105 figures, 3 tables. VIII, 226 pages
ISBN 3-540-08549-1

Contents: Primary Vestibular Neurons. – Central Vestibular Neurons. – Vestibulocerebellar Relationship. – Vestibuloocular Relationship.

Volume 3
J. T. Enright

The Timing of Sleep and Wakefulness

On the Substructure and Dynamics of the Circadian Pacemakers Underlying the Wake-Sleep Cycle

With a Foreword by E. Flory and an Appendix by J. Thorson
1980. 103 figures, 2 tables. XVIII, 263 pages
ISBN 3-540-09667-1

Contents: Introduction. – A Description of Activity-Rhythm Recordings and Their Implications. – The Pacemaker and its Precision. – A Class of Models for Mutual Entrainment of an Ensemble of Neurons. – A "Type Model" and its Behavior: Partial and Loose-Knit Mutual Entrainment. – Precision of Model Pacemakers. – Influences of Constant Light Intensity. – A Brief Detour: Further Thoughts About the Discriminator of the Models. – General Features of Entrainment: The Type Model. – Responses to Single Light Pulses. Part I: Nocturnal Rodents. – Responses to Single Light Pulses. Part II: Diurnal Birds. – Plasticity in Pacemaker Period: A Dynamic Memory. – Predictions from Coupled Stochastic Systems. – Further Predictions: A Modest Success and Two Problem Cases. – Morphology of the Models: Where is the Pacemaker? – A. Reprise and Synopsis: On the Advantages of Apparent Redundancy. – References. – Author Index. – Subject Index.

Springer-Verlag
Berlin
Heidelberg
New York

N. J. Strausfeld
Atlas of an Insect Brain
1976. 81 figures, partly coloured, 71 plates. XII, 214 pages
ISBN 3-540-07343-4

Contents: Introduction. – A Historical Commentary. – The Structure of Neuropil. – The Primary Compartments of the Brain. – The Coordinate System. – Some Quantitative Aspects of the Fly's Brain. – The Atlas: Sections through the Brain. – The Forms and Dispositions of Neurons in the Brain. – Appendix 1: Histological Methods. – Appendix 2: Dictionary of Terms. – References. – Subject Index.

This atlas is the first presentation of the main regions and pathways of an arthropod brain to combine both mass-staining of fibres and selective impregnation of neurons. It displays in detail the basic structures of the neuropils and schematizes them into a comprehensive and simple plan of sensory compartments and core neuropil. The main section of the book illustrates serial sections through the brain of the fly **Musca domestica** with reference to a coordinate system that relates covert structures to the head capsule. There follow detailed drawings of the forms and locations of Golgi-stained elements in the brain. The introductory chapters summarize the history of insect neuroanatomy, sketch the cellular constituents of neuropils, and outline the neuropil's basic organization. There is an appendix on histological methods applicable to insects and a multilingual glossary of terms relating to brain structures. The atlas is richly illustrated with 160 carefully prepared photographs and many beautiful drawings.

H. Stephan, W. K. Schwerdtfeger
The Brain of the Common Marmoset (Callithrix jacchus)
A Stereotaxic Atlas

1980. 5 figures, 3 tables, 73 plates. VI, 91 pages
ISBN 3-540-09782-1

Contents: Introduction. – Material. – Zero Coordinates. – Reference Coordinates. – Histological Procedures. – Shrinkage Factors. – Variability. – Standardization Proposals. – Photomicrographs for Atlas. – Presentation and Nomenclature. – References. – Index.

As anthropoid or simian primates, marmosets are close to man on the evolutionary scale. In contrast to other simians, the care and breeding of marmosets is relatively easy and their reproduction rate demonstrably higher. These characteristics make marmosets of great interest for primate research, especially as the supply of simians from the wild becomes increasingly limited.
This atlas presents 73 photomicrographs of successive serial sections from the brains of common marmosets. Taken at 0.5 mm intervals, the sections are stained for both cells and nerve fibers and shown on facing pages for comparison. Stereotactic coordinates are labelled in detail and methods for their standardization discussed.
The accuracy of the stereotactic descriptions provided here will greatly aid researchers using marmoset brains for neurophysiologic, neuroanatomic, neurochemical and behavioral investigations.

Springer-Verlag
Berlin
Heidelberg
New York